CORRELATION EQUATIONS
For Statistical Computations

CORRELATION EQUATIONS
For Statistical Computations

Aristarkh Konstantinovich Mitropol'skii
Academy of Wood Technology
Leningrad, USSR

Authorized translation from the Russian by
Edwin S. Spiegelthal

Springer Science+Business Media, LLC
1966

ISBN 978-1-4757-9876-0   ISBN 978-1-4757-9874-6 (eBook)
DOI 10.1007/978-1-4757-9874-6

The Russian text for this translation was prepared by the author in 1965
especially for this edition. It is based on Chapter 7 of his book *Tekhnika
Statisticheskikh Vychislenii,* published for the Physicomathematical
Engineering Library by the State Press for Physicomathematical Liter-
ature in Moscow in 1961.

Library of Congress Catalog Card Number 65-25246

PREFACE TO THE AMERICAN EDITION

This book presents methods for computing correlation equations. All the topics treated here are elucidated in terms of concrete examples, which have been chosen, for the most part, from the field of analysis of the mechanical properties of steel, wood, and other materials.

A necessary prerequisite for any study of correlation equations is some knowledge of the moments of random variables. In the Appendix, there is provided a brief treatment of moments, as well as a discussion of the simplest methods of computing them.

We have paid particular attention in this book to the techniques of computing correlation equations, and to the use of tables for alleviating the computational load. The mathematical bases of the methods used in setting up correlation equations are expounded in the books cited at the end of this volume.

A. M.

December 1965

Please note that the abbreviation lg is used in this book to designate the logarithm to base ten. Note further that the comma has been retained as the decimal point in tabular material.

v

CONTENTS

CHAPTER I. ORDINARY CORRELATION EQUATIONS

CHAPTER II. COMPUTATION OF CORRELATION EQUATIONS FOR SMALL SAMPLES

CHAPTER III. COMPUTING CORRELATION EQUATIONS BY MEANS OF CHEBYSHEV NUMBERS

CHAPTER IV. COMPUTING CORRELATION EQUATIONS BY THE METHOD OF SUMS

CHAPTER V. COMPUTATION OF CORRELATION EQUATIONS OF A NONPOLYNOMIAL TYPE

CHAPTER VI. MULTIPLE CORRELATION EQUATIONS

CHAPTER VII. DISTRIBUTION SURFACES

APPENDIX. MOMENTS

CHAPTER I

ORDINARY CORRELATION EQUATIONS

1. Method of Least Squares

The study of the relationships among random variables is one of the basic tasks of mathematical statistics. The simplest, and most important, such relationship is that of correlation among random variables, expressible by correlation equations. Correlation equations make it possible to compute the so-called probable value of one random variable as a function of the individual values of other random variables.

Calculation of probable values by means of correlation equations is of great practical importance, particularly in those cases in which a direct determination of the quantity under study is either accompanied by damage or even destruction of the work-piece, or else entails significant difficulty.

If the random variable X_2 is correlated with only one other variable, X_1, and conversely, then the equation expressing this relationship is called an ordinary correlation equation. If, now, each of the variables

$$X_1, X_2 ..., X_n$$

is correlated with several others of these variables, then the equations expressing these relationships are called multiple correlation equations.

We first consider ordinary correlation equations. Setting up a correlation equation reduces to the determination of the type of equation and then the calculation of the equation's coefficients. The type of correlation equation is a characteristic of the relationship between the random variables. In the majority of cases, the character of this relationship is such that, to express it, one uses a correlation equation in the form of a polynomial of one degree or another.

We shall now set up the correlation equation expressing the probable value, $\widetilde{X}_{(j_1)}|_1$, of the random variable X_2, as a function of the random variable X_1. For computational convenience, we shall initially attempt to express $r_{(j_1)}|_1$ approximately by means of an h'th-degree polynomial in the variable $\xi_{1(j_1)}$

$$r_{(j_1)|1}^{(h_1)} = \sum_{q_1=0}^{h_1} k_{q_1|}^{(h_1)} \xi_{1(j_1)}^{q_1} \tag{1}$$

In the approximate equality

$$r_{(j_1)|1} \approx r_{(j_1)|1}^{(h_1)} \tag{2}$$

we wish to signify the following. The coefficients $k_{q_1|}^{(h_1)}$ of equation (1) will be found by the method of least squares, i.e., from the condition that the sum of the squares of the differences between the left and right members of equation (2)

$$W_{h_1} = \sum_{j_1=1}^{k_1} p'_{j_1|} \left\{ r_{(j_1)|1} - \sum_{q_1=0}^{h_1} k_{q_1|}^{(h_1)} \xi_{1(j_1)}^{q_1} \right\}^2 \tag{3}$$

will be minimized.

To determine the coefficients $k_{q_1|.}^{(h_1)}$ it is necessary to find the partial derivatives of the function in (3) with respect to the $k_{f_1|.}^{(h_1)}$ and equate them to zero. We then have

$$\sum_{j_1=1}^{k_1} p_{j_1|.} \, \xi_{1(j_1)}^{f_1} \left\{ r_{(j_1)|1} - \sum_{q_1=0}^{h_1} k_{q_1|.}^{(h_1)} \, \xi_{1(j_1)}^{q_1} \right\} = 0. \tag{4}$$

By giving the exponent f_1 all integral values from 0 to h_1, we obtain a system of $h_1 + 1$ equations [cf., Appendix Section 3, equation (45)]:

$$r_{f_1|1} = \sum_{g_1=0}^{h_1} k_{g_1|.}^{(h_1)} \, r_{g_1+f_1|0} \qquad (f_1 = 0, h_1). \tag{5}$$

These are called normal equations.

Equations (5) are linear in the coefficients $k_{g_1|.}^{(h_1)}$. We thus wish to obtain a system of linear equations after differentiation of the function W_{h_1}, which explains our choice of the second power in expression (3).

To solve a system of normal equations, we use Cramer's rule, according to which

$$k_{g_1|.}^{(h_1)} = \frac{D_{g_1}^{h_1}}{D^{h_1}} \qquad (g_1 = \overline{0, h_1}), \tag{6}$$

where the determinant $D^{(h_1)}$ equals:

$$D^{(h_1)} = \begin{vmatrix} 1 & 0 & 1 & \cdots & r_{h_1|0} \\ 0 & 1 & r_{3|0} & \cdots & r_{h_1+1|0} \\ 1 & r_{3|0} & r_{4|0} & \cdots & r_{h_1+2|0} \\ \cdot & \cdot & \cdot & \cdots & \cdot \\ r_{h_1|0} & r_{h_1+1|0} & r_{h_1+2|0} & \cdots & r_{2h_1|0} \end{vmatrix}, \tag{7}$$

and the determinant $D_{g_1}^{h_1}$ is obtained from determinant $D^{(h_1)}$ by replacing the elements of its g_1'st column* by the standard moments

$$r_{0|1}, \; r_{1|1}, \; r_{2|1}, \; \ldots, \; r_{h_1|1},$$

which comprise the left-hand members of the normal equations in (5).

We call equation (1) an h_1'st-order correlation equation. We have

$$r_{(j_1)|1}^{(h_1)} = \sum_{g_1=0}^{h_1} \frac{D_{g_1}^{(h_1)}}{D^{(h_1)}} \, \xi_{1(j_1)}^{g_1}. \tag{8}$$

As an estimate of the degree to which the h_1'st-order correlation equation of (8) approximates the graph of $r_{(j_1)|1}$ as a function of $\xi_{|(j_1)}$, we choose the minimal value of W_{h_1} obtained from expression (3) when one substitutes in it the values just found for the coefficients $k_{g_1|.}^{(h_1)}$.

2. Chebyshev's Method

As we saw, there is no great difficulty in determining analytically the coefficients of correlation equation (1). However, in practice, the method of least squares turns out to be very inconvenient.

---

*The series of ordinal numbers here begins with zero.

2

In the majority of cases eventuating in computation, we lack information regarding the degree of that correlation equation which would provide a sufficiently accurate approximation to the aforementioned functional graph. It will happen, therefore, that the degree of the correlation equation must be raised.

But, as is clear from (1), the coefficients $k_{g_1}^{(h_1)}$ obtained for a correlation equation of degree h_1, are valueless when one goes to the correlation equation of degree $h_1 + 1$. Because of this, if the accuracy attained turns out to be inadequate, so that it is required — to increase that accuracy — to increase the degree of the correlation equation, then one must go through all the computational work all over again: set up, and solve, new normal equations, and compute, to estimate the accuracy, a new sum, analogous to the sum in (3).

To eliminate these inconveniences, P. L. Chebyshev suggested a special method for solving the problem of choosing a polynomial of some order or another. With Chebyshev's method, the higher-order terms of the equation are added successively, retaining all previously-obtained computational results and all estimates of the adequacy of the formulas obtained in each case.

The correlation equation to be developed by Chebyshev's method has the following general form:

$$r_{(j_1)|1}^{(h_1)} = \sum_{g_1=1}^{h_1} \frac{D_{g_1}^{(g_1)} D_{g_1}^{(g_1)*}}{D^{(g_1-1)} D^{(g_1)}}. \tag{9}$$

In this equation, $D^{(g_1)}$ and $D^{(g_1-1)}$ are the same determinants as in (7); $D_{g_1}^{(g_1)}$ is the determinant obtained from $D^{(g_1)}$ when the elements of its g_1'st column are replaced by the standard moments

$$r_{0|1}, \ r_{1|1}, \ r_{2|1}, \ ..., \ r_{g_1|1};$$

and, finally, $D_{g_1}^{(g_1)*}$ is the determinant obtained from $D^{(g_1)}$ by replacing the elements of its g_1'st column by the powers of the variable $\xi|_{(j_1)}$:

$$1, \ \xi_{1(j_1)}, \ \xi_{1(j_1)}^2, \ ..., \ \xi_{1(j_1)}^{g_1}.$$

Successively raising the order of the correlation equation, beginning with $h = 1$, we may present equation (9) in the form

$$r_{(j_1)|1}^{(h_1)} = r_{1|1}\xi_{1(j_1)} + \frac{b_1}{a_1}(\xi_{1(j_1)}^2 - r_{3|0}\xi_{1(j_1)} - 1) +$$

$$+ \frac{\begin{vmatrix} a_1 & b_1 \\ a_3 & b_2 \end{vmatrix}}{\begin{vmatrix} a_1 & a_3 \\ a_3 & a_5 \end{vmatrix}} \left\{ \xi_{1(j_1)}^3 - r_{4|0}\xi_{1(j_1)} - r_{3|0} - \frac{a_3}{a_1}(\xi_{1(j_1)}^2 - r_{3|0}\xi_{1(j_1)} - 1) \right\} +$$

$$+ \frac{\begin{vmatrix} a_1 & a_3 & b_1 \\ a_2 & a_3 & b_2 \\ a_4 & a_5 & b_3 \end{vmatrix}}{\begin{vmatrix} a_1 & a_3 & a_4 \\ a_2 & a_3 & a_5 \\ a_4 & a_5 & a_6 \end{vmatrix}} \left\{ \xi_{1(j_1)}^4 - r_{5|0}\xi_{1(j_1)} - r_{4|0} - \frac{\begin{vmatrix} a_1 & a_4 \\ a_2 & a_5 \\ a_1 & a_2 \\ a_3 & a_3 \end{vmatrix}}{} (\xi_{1(j_1)}^3 - r_{4|0}\xi_{1(j_1)} - r_{3|0}) - \right.$$

$$\left. - \frac{\begin{vmatrix} a_4 & a_2 \\ a_5 & a_3 \\ a_1 & a_2 \\ a_2 & a_3 \end{vmatrix}}{} (\xi_{1(j_1)}^2 - r_{3|0}\xi_{1(j_1)} - 1) \right\} + ... \tag{10}$$

In this equation

$$
\left.\begin{aligned}
a_1 &= r_{4|0} - r_{3|0}^2 - 1, \\
a_2 &= r_{5|0} - r_{4|0}r_{3|0} - r_{3|0}, \\
a_3 &= r_{6|0} - r_{4|0}^2 - r_{3|0}^2, \\
a_4 &= r_{6|0} - r_{5|0}r_{3|0} - r_{4|0}, \\
a_5 &= r_{7|0} - r_{5|0}r_{4|0} - r_{4|0}r_{3|0}, \\
a_6 &= r_{8|0} - r_{5|0}^2 - r_{4|0}^2;
\end{aligned}\right\}
\tag{11}
$$

$$
\left.\begin{aligned}
b_1 &= r_{2|1} - r_{3|0}r_{1|1}, \\
b_2 &= r_{3|1} - r_{4|0}r_{1|1}, \\
b_3 &= r_{4|1} - r_{5|0}r_{1|1}.
\end{aligned}\right\}
\tag{12}
$$

$$
\left.\begin{aligned}
\begin{vmatrix} a_1 & a_2 \\ a_2 & a_3 \end{vmatrix} &= a_1a_3 - a_2^2, \\[4pt]
\begin{vmatrix} a_1 & b_1 \\ a_3 & b_2 \end{vmatrix} &= a_1b_2 - a_3b_1, \\[4pt]
\begin{vmatrix} a_1 & a_4 \\ a_2 & a_5 \end{vmatrix} &= a_1a_5 - a_2a_4, \\[4pt]
\begin{vmatrix} a_4 & a_2 \\ a_5 & a_3 \end{vmatrix} &= a_3a_4 - a_2a_5 = - \begin{vmatrix} a_2 & a_4 \\ a_3 & a_5 \end{vmatrix}, \\[6pt]
\begin{vmatrix} a_1 & a_2 & a_4 \\ a_2 & a_3 & a_5 \\ a_4 & a_5 & a_6 \end{vmatrix} &= a_4 \begin{vmatrix} a_2 & a_4 \\ a_3 & a_5 \end{vmatrix} - a_5 \begin{vmatrix} a_1 & a_4 \\ a_2 & a_5 \end{vmatrix} + a_6 \begin{vmatrix} a_1 & a_2 \\ a_2 & a_3 \end{vmatrix}, \\[6pt]
\begin{vmatrix} a_1 & a_2 & b_1 \\ a_2 & a_3 & b_2 \\ a_4 & a_5 & b_3 \end{vmatrix} &= b_1 \begin{vmatrix} a_2 & a_4 \\ a_3 & a_5 \end{vmatrix} - b_2 \begin{vmatrix} a_1 & a_4 \\ a_2 & a_5 \end{vmatrix} + b_3 \begin{vmatrix} a_1 & a_2 \\ a_2 & a_3 \end{vmatrix}.
\end{aligned}\right\}
\tag{13}
$$

The conversion from the approximate conditional standard moments, $r_{(j_1)}^{(h_1)}|_1$, to the probable values is accomplished by means of the formula

$$
\tilde{X}_{(j_1)|1} = \bar{X}_2 + r_{(j_1)|1}^{(h_1)}\,\bar{\sigma}_2
\tag{14}
$$

[cf., Appendix, Section 3, equation (47)].

By using (10), we can make the transition from a first-order correlation equation to correlation equations of ever higher orders. To indicate at which order of correlation equation one should stop, there is the correlation equation criterion

$$
\zeta_{h_1} = \zeta_{h_1-1} - \frac{D_{h_1}^{(h_1)2}}{D^{(h_1-1)}D^{(h_1)}}
\tag{15}
$$

with its standard error

$$
\sigma_{\zeta_{h_1}} = \sqrt{\frac{\zeta_{h_1}}{n}},
\tag{16}
$$

with

$$
\zeta_1 = \sum_{j_1=1}^{k_1} p_{j_1}.\, r_{(j_1)|1}^2 - r^2|_1.
\tag{17}
$$

4

If the magnitude of the criterion ζ_{h_1} turns out to be sufficiently small in comparison with its standard error, $\sigma \zeta_{h_1}$, then we may halt at the correlation equation of order h_1.

Equation (14), approximately expressing the conditional mean value, $\overline{X}_{(j_1)}|_1$, may even be used for an approximate representation of the $X_{2(j_2)}$ themselves. The standard error, $\sigma_{2.1}^{(h_1)}$, in defining the value of $X_{2(j_2)}$ by formula (14) or, simply, the standard error of the equation, is computed by the formula

$$\frac{\sigma_{2.1}^{(h_1)2}}{\sigma_2^2} = 1 - \sum_{g_1=1}^{h_1} \frac{D_{g_1}^{(g_1)2}}{D^{(g_1-1)} D^{(g_1)}} \tag{18}$$

Were we to stop with the first term of equation (10), we would obtain a linear correlation equation

$$r_{(j_1)1}^{(1)} = r_{1|1} \xi_{1(j_1)}, \tag{19}$$

whose standard error is

$$\sigma_{2.1}^{(1)} = \sigma_2 \sqrt{1 - r_{1|1}^2}. \tag{20}$$

As the estimate of the first-order correlation equation, we have the linearity criterion

$$\zeta_1 = \eta_{21}^2 - r_{1|1}^2 \tag{21}$$

with standard error

$$\sigma_{\zeta_1} = \sqrt{\frac{\zeta_1}{n}}. \tag{22}$$

Taking (14) into account, we may present the correlation equation of (19) in the form

$$\tilde{X}_{(j_1)|1} = \overline{X}_2 + r_{1|1} \frac{\overline{\sigma}_2}{\sigma_1} (X_{1(j_1)} - \overline{X}_1). \tag{23}$$

Similarly,

$$\tilde{X}_{1|(j_2)} = \overline{X}_1 + r_{1|1} \frac{\overline{\sigma}_1}{\sigma_2} (X_{2(j_2)} - \overline{X}_2). \tag{24}$$

Joining to the first term of equation (10) its second term, we obtain the second-order correlation equation

$$r_{(j_1)|1}^{(2)} = r_{1|1} \xi_{1(j_1)} + \frac{b_1}{a_1} (\xi_{1(j_1)}^2 - r_{3|0} \xi_{1(j_1)} - 1) \tag{25}$$

with standard error

$$\sigma_{2.1}^{(2)} = \sigma_2 \sqrt{1 - r_{1|1}^2 - \frac{b_1^2}{a_1}} \tag{26}$$

For estimating this equation, we have the quadratic criterion

$$\zeta_2 = \zeta_1 - \frac{b_1^2}{a_1} \tag{27}$$

with standard error

$$\sigma_{\zeta_2} = \sqrt{\frac{\zeta_2}{n}}. \tag{28}$$

5

TABLE 1

$E \cdot 10^{-3}$, kg/cm² $\gamma \cdot 10^3$, g/cm³	60	74	88	102	116	130	144	158	172	186	200	Σ
325	1	3	2									6
355	1	1	9	6	1							18
385		1	8	20	8							37
415			1	11	42	14						68
445				7	26	46	11					90
475					9	31	36	3				79
505					1	9	23	19	2			54
535							8	15	6	2		31
565								1	4	6	1	12
595									1	2	1	4
625											1	1
Σ	2	5	20	44	87	100	79	42	16	4	1	400

By adding the third term to the first two terms of equation (10), we obtain the third-order correlation equation

$$r_{(j_1)\|1}^{(3)} = r_{1\|1}\xi_{1(j_1)} + \frac{b_1}{a_1}\left(\xi_{1(j_1)}^2 - r_{3|0}\xi_{1(j_1)} - 1\right) +$$

$$+ \frac{\begin{vmatrix} a_1 & b_1 \\ a_2 & b_2 \end{vmatrix}}{\begin{vmatrix} a_1 & a_2 \\ a_2 & a_3 \end{vmatrix}}\left\{\xi_{1(j_1)}^3 - r_{4|0}\xi_{1(j_1)} - r_{3|0} - \frac{a_2}{a_1}\left(\xi_{1(j_1)}^2 - r_{3|0}\xi_{1(j_1)} - 1\right)\right\}. \tag{29}$$

The standard error of this correlation equation equals

$$\sigma_{2.1}^{(3)} = \sigma_2 \sqrt{1 - r_{1\|1}^2 - \frac{b_1^2}{a_1} - \frac{\begin{vmatrix} a_1 & b_1 \\ a_2 & b_2 \end{vmatrix}^2}{a_1\begin{vmatrix} a_1 & a_2 \\ a_2 & a_3 \end{vmatrix}}}. \tag{30}$$

TABLE 2

$X_{1(j_1)}$	$\overline{X}_{(j_1)\|1}$	$\tilde{X}_{(j_1)\|1}$
0,325	79 138	81 404
0,355	91 896	92 351
0,385	101 244	103 298
0,415	116 290	114 245
0,445	125 492	125 192
0,475	135 852	136 139
0,505	147 108	147 086
0,535	158 910	158 033
0,565	166 162	168 980
0,595	172 000	179 927
0,625	200 000	190 874
Σ	1494 092	1497 529

As the figure of merit of the third-order equation, we compute the cubic criterion

$$\zeta_3 = \zeta_2 - \frac{\begin{vmatrix} a_1 & b_1 \\ a_2 & b_2 \end{vmatrix}^2}{a_1 \begin{vmatrix} a_1 & a_2 \\ a_2 & a_3 \end{vmatrix}} \tag{31}$$

with standard error

$$\sigma_{\zeta_3} = \sqrt{\frac{\zeta_3}{n}}. \tag{32}$$

By thus proceeding, we can eventually obtain a correlation equation, of some order or another, which will express, with the requisite accuracy, the functional dependence of $\eta_{(j_1)}\|1$ on $\xi\|_{(j_1)}$, or we will obtain equation (14), giving the approximate conditional mean, $\overline{X}_{(j_1)}\|1$; this latter equation may even be used for computing the values of the random variable X_2 from the known values of variable X_1.

3. First-Order Correlation Equations

We shall consider the successive computations required in setting up a first-order correlation equation by working out the example of the correlation equation which expresses the dependence of the modulus of elasticity X_2 (E, kg/cm^2), with compression along the grain, on the weight-by-volume X_1 (γ, g/cm^3) of spruce with 10% moisture content (Table 1).

The salient statistics from this table turn out to be the following:

$$m_{1|0} = +0.335, \quad m_{0|1} = -0.082, \quad \overline{X}_1 = 0.455 \text{ g/cm}^3,$$
$$\overline{X}_2 = 128852 \text{ kg/cm}^2, \quad \sigma_1 = 1.824, \quad \sigma_2 = 1.620, \quad r_{1|1} = +0.880, \quad \eta_{21} = 0.882.$$

From our computations, we find that

$$\zeta_1 = 0.004 \pm 0.003.$$

Thus, we can rest content with a first-order correlation equation.

By substituting the statistics we have obtained into equation (23), we obtain the following correlation equation expressing the relationship of interest to us

$$\tilde{X}_{(j_1)\|1} = -37188 + 364990 \; X_{1(j_1)} \tag{33}$$

with standard error

$$\overline{\sigma}_{2.1} = 10780 \text{ kg/cm}^2.$$

In Table 2 we give the probable values, computed on the basis of (33), together with the observed conditional means, $\overline{X}_{(j_1)}\|1$, for the given values, $X_{1(j_1)}$, of the weight-by-volume of spruce.

The tabular data are presented on Fig. 1 (the circles are observed conditional means, the solid line gives the probable values). The same figure shows, by dotted lines, the zone of variability, defined by the standard error of the correlation equation.

In comparing the probable values with the corresponding means, we note a virtually exact coincidence between them.

TABLE 3

X_1, g \ X_2, %	5	5,5	6	6,5	7	7,5	8	8,5	9	9,5	10	10,5	Σ
125		1											1
150	1	2	2										5
175	1	3	6	7	5	1							23
200	1	2	9	16	19	8	3	1					59
225		1	5	21	34	42	24	7					134
250			2	12	29	58	66	32	4				203
275				4	12	44	73	56	19	4			212
300					5	16	42	60	43	11	2		179
325					1	4	17	37	34	14	6	1	114
350						1	3	10	14	17	5	1	51
375								2	4	5	3	1	15
400									1	1	1		3
425										1			1
Σ	3	9	24	60	105	174	228	205	119	53	17	3	1000

4. Second-Order Correlation Equations

We now consider the correlation between strength, X_1 (g), and stretching, X_2 (%), of cotton thread, (cf., Table 3*), setting ourselves the problem of deriving the correlation equation expressing the functional dependence of stretching on the thread's strength.

The statistics for Table 3, and the coefficients, a_1 and b_1, of the correlation equations, are given in Table 4.

In attempting now to ascertain the necessary order for the required correlation equation, we find

$$\zeta_1 = 0.005 \pm 0.002,$$
$$\zeta_2 = 0.$$

---

*Data according to E. A. Sankov.

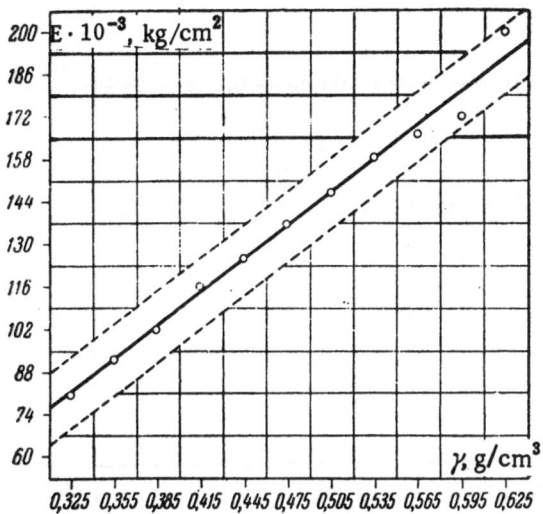

Fig. 1. Spruce's modulus of elasticity, with compression along the grain, as function of the weight-by-volume, with wood of 10% moisture content.

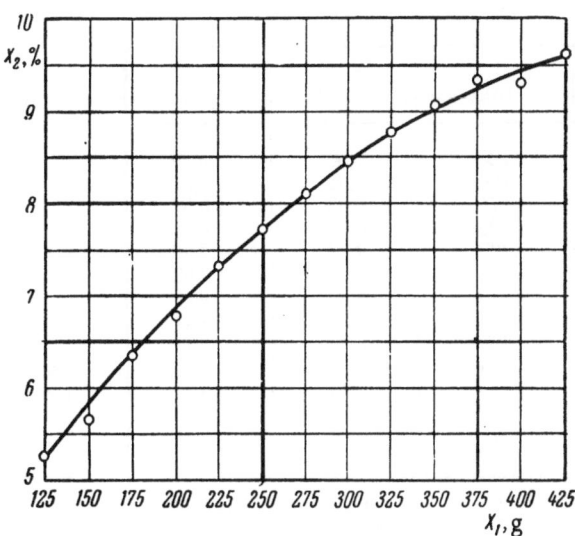

Fig. 2. Functional dependence of stretching, X_2, on strength, X_1, of cotton threads.

Consequently, the dependence of stretching on thread strength can, apparently, be sufficiently well expressed by a second-order correlation equation. This correlation equation turned out to be the following:

$$r^{(2)-}_{(j_1)|1} = 0.751\xi_{1(j_1)} + \frac{-0.101}{1.910}(\xi^2_{1(j_1)} - 0.017\xi_{1(j_1)} - 1) = 0.053 + 0.752\xi_{1(j_1)} - 0.053\xi^2_{1(j_1)}$$

TABLE 4

Statistics	X_1	X_2
m_1	−0,130	−0,038
m_2	3,244	3,356
m_3	−1,162	−1,892
m_4	30,580	35,000
μ_2	3,227	3,357
μ_3	+0,0 9	
μ_4	30,304	
r_3	+0,017	
r_4	2,910	
$\bar{X}$	271,75	7,98
σ	1,796	1,832
$\bar{\sigma}$	44,90	0,916
v	16,52	11,48

$m_{1|1} = +2,476,$ $m_{2|1} = −1,288,$

$\mu_{1|1} = +2,471,$ $\mu_{2|1} = −0,522,$

$r_{1|1} = +0,751,$ $r_{2|1} = −0,088,$

$\eta_{2 \cdot 1} = 0,754,$

$a_1 = 1,910,$ $b_1 = −0,101.$

with standard error

$$\sigma^{(2)}_{2.1} = 1.203.$$

To check the correctness of the equation we have obtained, we calculate it by another method, namely, by formula (8).

It is easily seen that

$$D^{(2)} = a_1 = 1.910,$$
$$D^{(2)}_0 = -b_1 = +0.101,$$
$$D^{(2)}_1 = r_{1|1}a_1 - r_{3|0}b_1 = 1.436,$$
$$D^{(2)}_2 = b_1 = -0.101.$$

Consequently,

$$r^{(2)}_{(j_1)|1} = \frac{0.101}{1.910} + \frac{1.436}{1.910}\xi_{1(j_1)} - \frac{0.101}{1.910}\xi^2_{1(j_1)}$$

$$= 0.053 + 0.752\xi_{1(j_1)} - 0.053\xi^2_{1(j_1)},$$

as we established above.

TABLE 5

$X_{1(j_1)}$	$\overline{X}_{(j_1)|1}$	$\tilde{X}_{(j_1)|1}$
125	5,50	5,26
150	5,60	5,81
175	6,33	6,32
200	6,77	6,81
225	7,29	7,26
250	7,70	7,68
275	8,06	8,08
300	8,45	8,44
325	8,75	8,77
350	9,11	9,08
375	9,40	9,35
400	9,50	9,59
425	9,50	9,81
Σ	(101,96)	(102,26)

The equation we have found may be put in the form of (14):

$$\tilde{X}_{(j_1)|1}^{(2)} = 7.98 + 0.916 \left\{ 0.053 + 0.752 \left(\frac{X_{1(j_1)} - 271.75}{44.90} \right) \right.$$
$$\left. - 0.053 \left(\frac{X_{1(j_1)} - 271.75}{44.90} \right)^2 \right\}$$
$$= 2.09 + 0.0284 X_{1(j_1)} - 0.0000241 X_{1(j_1)}^2.$$

The probable values of stretching, $\tilde{X}_{(j_1)|1}$ for given values, $X_{1(j_1)}$, of thread strength, computed from this equation, together with the observed conditional means, $\overline{X}_{(j_1)|1}$, are given both in Table 5 and in Fig. 2.

5. Third-Order Correlation Equations

As an example of third-order correlation equations, we may use the correlation equation expressing yield strength to rupture, $\dot{X}_2$ (σ_B, kg/mm²), of stressed steel on toughness, X_1. (a_k, kgm/cm²) (Table 6).

The necessary statistics for this case are provided in Table 7.

TABLE 6

a_k, kgm/cm² \ σ_B, kg/mm²	48	50	52	54	56	58	60	62	64	Σ
4,5								1	1	2
5,5					2	3	3	2		10
6,5			5	7	21	27	21	10	1	92
7,5	1	4	11	28	48	43	29	5		169
8,5		4	22	60	63	38	23	1	3	214
9,5		5	15	32	47	26	6	1		132
10,5	1	2	14	37	28	12	6	3		103
11,5	1	2	4	22	16	6	2			53
12,5			4	8	3		1			16
Σ	3	17	75	194	228	155	91	23	5	791

TABLE 7. Various Statistics for Table 6

$m_{110} = +0,2149$		
$m_{210} = 2,4475$	$\mu_{210} = 2,4013$	
$m_{310} = +2,6650$	$\mu_{310} = 1,1069$	$r_{310} = 0,2975$
$m_{410} = 16,6018$	$\mu_{410} = 14,9831$	$r_{410} = 2,5984$
$m_{510} = +31,7320$	$\mu_{510} = 14,8839$	$r_{510} = 1,6620$
$m_{610} = 167,4286$	$\mu_{610} = 137,5674$	$r_{610} = 9,9133$
$m_{710} = +410,6043$	$\mu_{710} = 183,9507$	$r_{710} = 8,5543$
$m_{810} = 2077,3856$	$\mu_{810} = 1572,8545$	$r_{810} = 47,2008$

$$\overline{X}_1 = 8,715 \pm 0,055 \text{ kgm/cm}^2 \quad \sigma_1 = 1,5496 \pm 0,0390,$$
$$c_1 = 1 \text{ kgm/cm}^2, \quad v_1 = 17,78 \pm 046\%.$$

$$m_{011} = +0,0240, \quad m_{012} = 1,8976, \quad \mu_{012} = 1,8970,$$
$$c_2 = 2 \text{ kg/mm}^2 \quad \overline{X}_2 = 56,048 \pm 0,098 \text{ kg/mm}^2$$
$$\sigma_2 = 1,3773 \pm 0,0346, \quad v_2 = 4,91 \pm 0,12\%.$$

$m_{111} = -0,7092$	$\mu_{111} = -0,7144$	$r_{111} = -0,3347$
$m_{211} = -0,0721$	$\mu_{211} = +0,1762$	$r_{211} = +0,0533$
$m_{311} = -4,6384$	$\mu_{311} = -4,7170$	$r_{311} = -0,9204$
$m_{411} = -2,8938$	$\mu_{411} = +0,7469$	$r_{411} = +0,0934$

$a_1 = 1,5099,$	$a_2 = 0,5915,$	$a_3 = 3,0731$
$a_4 = 6,8205,$	$a_5 = 3,4627,$	$a_6 = 37,6869$
$b_1 = +0,1529,$	$b_2 = -0,0507,$	$b_3 = +0,6497$

$$\begin{vmatrix} a_1 & a_2 \\ a_2 & a_3 \end{vmatrix} = 4,2902, \qquad \begin{vmatrix} a_1 & b_1 \\ a_2 & b_2 \end{vmatrix} = -0,1670,$$

$$\begin{vmatrix} a_1 & a_2 & a_4 \\ a_2 & a_3 & a_5 \\ a_4 & a_5 & a_6 \end{vmatrix} = 28,5603, \qquad \begin{vmatrix} a_1 & a_2 & b_1 \\ a_2 & a_3 & b_2 \\ a_4 & a_5 & b_3 \end{vmatrix} = -0,0436.$$

Since, in this example of the dependence of yield strength on toughness of axial steel, we have

$$\eta_{21} = 0,3644 \pm 0,0318,$$
$$r_{1|1} = -0,3347 \pm 0,0316,$$

the linearity criterion equals

$$\zeta_1 = 0,0209 \pm 0,0052.$$

Consequently,

$$\frac{\zeta_1}{\sigma_{\zeta_1}} = 4,02;$$

this indicates that the dependence of yield strength on toughness cannot be adequately represented by a first-order equation.

The quadratic criterion is

$$\zeta_2 = 0,0054 \pm 0,0026,$$

with

$$\frac{\zeta_2}{\sigma_{\zeta_2}} = 2,07.$$

The cubic criterion is

$$\zeta_3 = 0,0011 \pm 0,0012,$$

TABLE 8. Computation of $r_{(j_1)|1}^{(3)}$ by Equation (34)

$X_{1(j_1)}$	$x_{1(j_1)} - m_{1|0}$	$\xi_{1(j_1)}$	$\xi_{1(j_1)}^2$	$\xi_{1(j_1)}^3$	$-0{,}1049 - 0{,}2682\,\xi_{1(j_1)} + 0{,}1165\,\xi_{1(j_1)}^2 - 0{,}0389\,\xi_{1(j_1)}^3$	$r_{(j_1)|1}^{(3)}$
4,5	− 4,2149	− 2,7200	7,3984	− 20,1236	− 0,1049 + 0,7295 + 0,8619 + 0,7828	+ 2,2693
5,5	− 3,2149	− 2,0747	4,3044	− 8,9303	− 0,1049 + 0,5564 + 0,5015 + 0,3474	+ 1,3004
6,5	− 2,2149	− 1,4293	2,0429	− 2,9199	− 0,1049 + 0,3833 + 0,2380 + 0,1136	+ 0,6300
7,5	− 1,2149	− 0,7840	0,6147	− 0,4819	− 0,1049 + 0,2103 + 0,0716 + 0,0187	+ 0,1957
8,5	− 0,2149	− 0,1387	0,0192	− 0,0027	− 0,1049 + 0,0372 + 0,0022 + 0,0001	− 0,0654
9,5	+ 0,7851	+ 0,5066	0,2566	+ 0,1300	− 0,1049 − 0,1359 + 0,0299 − 0,0051	− 0,2160
10,5	+ 1,7851	+ 1,1520	1,3271	+ 1,5288	− 0,1049 − 0,3090 + 0,1546 − 0,0595	− 0,3188
11,5	+ 2,7851	+ 1,7973	3,2303	+ 5,8058	− 0,1049 − 0,4820 + 0,3763 − 0,2258	− 0,4364
12,5	+ 3,7851	+ 2,4426	5,9663	+ 14,5733	− 0,1049 − 0,6551 + 0,6951 − 0,5669	− 0,6318

TABLE 9

$X_{1(j_1)}$	$\overline{X}_{(j_1)|1}$	$\tilde{X}_{(j_1)|1}$
4,5	63,00	62,30
5,5	59,00	59,63
6,5	57,87	57,79
7,5	56,59	56,59
8,5	55,84	55,87
9,5	55,45	55,46
10,5	55,18	55,17
11,5	54,87	54,85
12,5	54,25	54,31
Σ	(512,05)	(511,97)

with

$$\frac{\zeta_3}{\sigma_{\zeta_3}} = 0{,}92.$$

Finally, the quantity ζ_4 is only 0.00001 less than ζ_3, i.e., is virtually equal to ζ_3.

We thus conclude that a third-order equation, but no higher, is necessary for expressing the relationship of yield strength to toughness of stressed axial steel.

By substituting our statistics in (29), we obtain the third-order correlation equation in the following form:

$$r_{(j_1)|1}^{(3)} = \frac{\begin{array}{l} -0{,}3347\,\xi_{1(j_1)} - \\ -0{,}1013 - 0{,}0301\,\xi_{1(j_1)} + 0{,}1013\,\xi_{1(j_1)}^2 + \\ +0{,}0116 + 0{,}1011\,\xi_{1(j_1)} \qquad\qquad -0{,}0389\,\xi_{1(j_1)}^3 - \\ -0{,}0152 - 0{,}0045\,\xi_{1(j_1)} + 0{,}0152\,\xi_{1(j_1)}^3 \end{array}}{-0{,}1049 - 0{,}2682\,\xi_{1(j_1)} + 0{,}1165\,\xi_{1(j_1)}^2 - 0{,}0389\,\xi_{1(j_1)}^3.} \tag{34}$$

The standard error of this correlation equation is

$$\sigma_{2.1}^{(3)} = 1.1958.$$

TABLE 10. Computation of $\tilde{X}^{(3)}_{(j_1)|1}$ by Equation (38)

$X_{1(j_1)}$	$X^2_{1(j_1)}$	$X^3_{1(j_1)}$	$89{,}13 - 9{,}3684\,X_{1(j_1)} + 0{,}8866\,X^2_{1(j_1)} - 0{,}0288\,X^3_{1(j_1)}$	$\tilde{X}^{(3)}_{(j_1)|1}$
4,5	20,25	91,125	$89{,}13 - 42{,}1578 + 17{,}9536 - 2{,}6244$	62,30
5,5	30,25	166,375	$89{,}13 - 51{,}5262 + 26{,}8196 - 4{,}7916$	59,63
6,5	42,25	274,625	$89{,}13 - 60{,}8946 + 37{,}4588 - 7{,}9092$	57,79
7,5	56,25	421,875	$89{,}13 - 70{,}2630 + 49{,}8712 - 12{,}1500$	56,59
8,5	72,25	614,125	$89{,}13 - 79{,}6314 + 64{,}0568 - 17{,}6868$	55,87
9,5	90,25	857,375	$89{,}13 - 88{,}9998 + 80{,}0156 - 24{,}6924$	55,45
10,5	110,25	1157,625	$89{,}13 - 98{,}3682 + 97{,}7476 - 33{,}3396$	55,17
11,5	132,25	1520,875	$89{,}13 - 107{,}7366 + 117{,}2528 - 43{,}8012$	54,84
12,5	156,25	1953,125	$89{,}13 - 117{,}1050 + 138{,}5312 - 56{,}2500$	54,31

To verify the correctness of the correlation equation just found, we recompute it from formula (8). The determinants which enter into this formula equal

$$D^{(3)} = \quad 4{,}2902,$$
$$D^{(3)}_0 = -0{,}4502,$$
$$D^{(3)}_1 = -1{,}1507,$$
$$D^{(3)}_2 = +0{,}4999,$$
$$D^{(3)}_3 = -0{,}1670.$$

Consequently,

$$r^{(3)}_{(j_1)|1} = \frac{1}{4{,}2902}\left(-0{,}4502 - 1{,}1507\,\xi_{1(j_1)} + 0{,}4999\,\xi^2_{1(j_1)} - 0{,}1670\,\xi^3_{1(j_1)}\right) = -0{,}1049 - 0{,}2682\,\xi_{1(j_1)} + 0{,}1165\,\xi^2_{1(j_1)} - 0{,}0389\,\xi^3_{1(j_1)}. \tag{35}$$

Once having found some value or other for the quantity

$$\xi_{1(j_1)} = \frac{x_{1(j_1)} - m_{1|0}}{\sigma_1} \tag{36}$$

we then find the corresponding value of $r^{(3)}_{(j_1)|1}$. The calculations are laid out in Table 8.

From the values of $r^{(3)}_{(j_1)|1}$, we then find the probable values of yield strength for stressed axial steel. To do this, we employ formula (14), which in the present case, takes the form

$$\tilde{X}^{(3)}_{(j_1)|1} = 56{,}048 + 2{,}7546\,r^{(3)}_{(j_1)|1} \tag{37}$$

with standard error

$$\sigma^{(3)}_{2.1} = 2{,}3916.$$

The results of the computations are given in Table 9. To check the probable values we have found, we recompute them by another method.

By substituting in (37) the expression for $r^{(3)}_{(j_1)|1}$ from (34), using (36) to substitute for $\xi_{1(j_1)}$, we obtain

$$\tilde{X}^{(3)}_{(j_1)|1} = 56.048 + 2.7546 \left\{ -0.1049 - 0.2682 \left(\frac{X_{1(j_1)} - 8.7149}{1.5496} \right) + \right.$$

$$\left. + 0.1165 \left(\frac{X_{1(j_1)} - 8.7149}{1.5496} \right)^2 - 0.0389 \left(\frac{X_{1(j_1)} - 8.7149}{1.5496} \right)^3 \right\},$$

or

$$\tilde{X}^{(3)}_{(j_1)|1} = 89.13 - 9.3684\, X_{1(j_1)} + 0.8864\, X^2_{1(j_1)} - 0.0288\, X^3_{1(j_1)}. \tag{38}$$

The computed probable values, obtained from this equation, are given in Table 10.

Figure 3 shows the observed conditional means of yield strength of stressed axial steel (circles) and the probable values (solid line). The same figure shows the variance zone, bounded by dotted lines at a distance, on each side of the solid line, equal to the standard error of the correlation equation.

Comparison of the probable values, as computed from the correlation equation, with the observed means, shows that the third-order correlation equation sufficiently well expresses the dependence of yield strength of stressed axial steel on toughness.

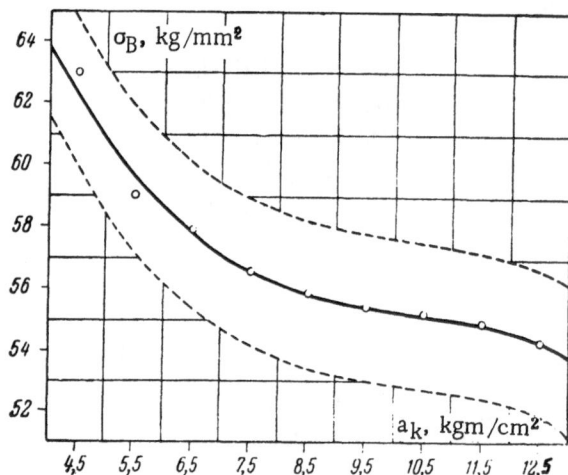

Fig. 3. Dependence of yield strength of axial steel on toughness.

CHAPTER II

COMPUTATION OF CORRELATION EQUATIONS FOR SMALL SAMPLES

1. Computing First-Order Correlation Equations

When the sample is small, the observed values of the random variables are not grouped. Therefore, in computing correlation equations here, one has to deal with nonuniformly spaced values of the quantities being investigated. Due to this, the usual methods of finding the requisite moments turn out to be inadequate, and it becomes necessary to develop techniques which would be applicable to small samples. We shall consider these techniques as applied in various examples.

We shall set up the correlation equation expressing the dependence of solubility* of sodium nitrate, X_2, on the variation of temperature, X_1 (Table 11), an example used by D. I. Mendeleev.†

By graphing these data (Fig. 4), we see that the solubility of the sodium nitrate varies practically proportionally to the changes in temperature. Consequently, the relationship being investigated may apparently be expressed by a first-order equation.

We have presented the statistics needed for setting up the linear correlation equation in Table 12.

In column (1) of our layout, we have listed the ordinal numbers of the trials. Columns (2) and (3) give the observed values of the quantities being studied. The sums of these values, divided by the sample size, give the means. In our case,

$$\overline{X}_1 = \frac{234}{9} = 26, \quad \overline{X}_2 = \frac{811.3}{9} = 90.1.$$

Columns (4) and (5) contain the deviations of each of the observed quantities from its respective mean. The contents of the remaining columns are self-explanatory.

On the basis of the totals of columns (6), (7), and (8), we have·

$$\mu_{2|0} = \frac{4060}{9} = 451.11, \quad \mu_{0|2} = \frac{3084.00}{9} = 342.67,$$
$$\mu_{1|1} = \frac{3534.8}{9} = 392.76.$$

TABLE 11

X_1°C	0	4	10	15	21	29	36	51	68
X_2	66.7	71.0	76.3	80.6	85.7	92.9	99.4	113.6	125.1

*Solubility is defined as the quantity of the material capable of saturating 100 parts of water.

†D. I. Mendeleev, Basic Chemistry, Volume I, 12th Edition, Leningrad, 1934, p. 389.

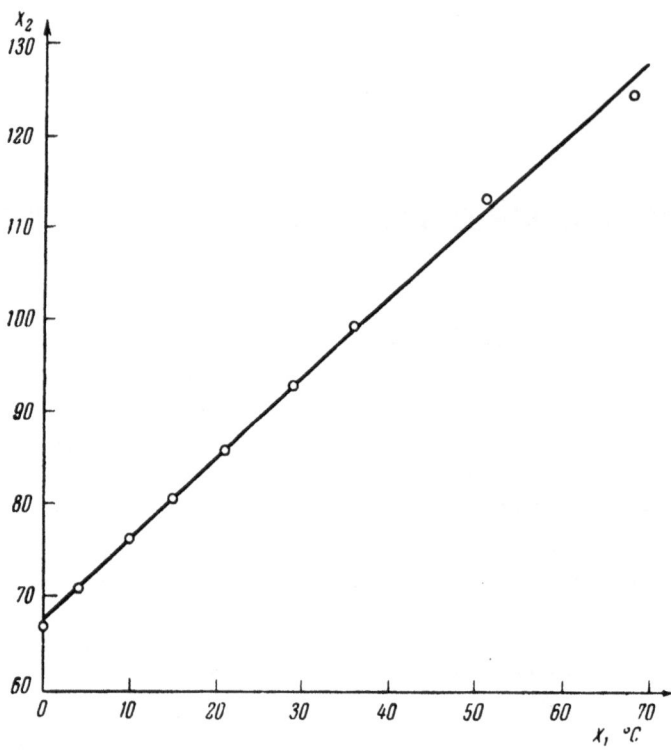

Fig. 4. Dependence of the solubility, X_2, of sodium nitrate on
variations in temperature, X_1.

Consequently,

$$\sigma_1 = \sqrt{451.11} = 21.239, \quad \sigma_2 = \sqrt{342.67} = 18.512,$$
$$r_{1|1} = \frac{392.76}{21.239 \cdot 18.512} = \frac{392.76}{393.18} = +0.999.$$

The last two columns of our layout serve for checking the computations. By dividing the sum of column (10)
by the number of trials, we obtain empirically the second central moment of the differences of the two quantities

$$\mu_{2\,(d)} = \frac{74.40}{9} = 8.27.$$

By then using the formula

$$\mu_{2\,(d)} = \mu_{2|0} + \mu_{0|2} - 2\mu_{1|1}$$

we find

$$451.11 + 342.67 - 785.52 = 8.26.$$

Thus, our computations were correct.

In order to determine the order of the correlation equation, we compute the linearity criterion, together
with its standard error:

$$\zeta_1 = 1 - 0.999^2 = 0.002,$$
$$\sigma_{\zeta_1} = \sqrt{\frac{0.002}{9}} = 0.014.$$

The ratio of these quantities

$$\frac{\zeta_1}{\sigma_{\zeta_1}} = 0.14.$$

TABLE 12. Layout of the Computation of Moments for Nonuniformly
Spaced Variable Values

№	X_1	X_2	x_1	x_2	x_1^2	x_2^2	$x_1 x_2$	$x_1 - x_2$	$(x_1 - x_2)^2$
(1)	(2)	(3)	(4)	(5)	(6)	(7)	(8)	(9)	(10)
1	0	66,7	− 26	− 23,4	676	547,56	608,4	− 2,6	6,76
2	4	71,0	− 22	− 19,1	484	364,81	420,2	− 2,9	8,41
3	10	76,3	− 16	− 13,8	256	190,44	220,8	− 2,2	4,84
4	15	80,6	− 11	− 9,5	121	90,25	104,5	− 1,5	2,25
5	21	85,7	− 5	− 4,4	25	9,36	22,0	− 0,6	0,36
6	29	92,9	+ 3	+ 2,8	9	7,84	8,4	+ 0,2	0,04
7	36	99,4	+ 10	+ 9,3	100	86,49	93,0	+ 0,7	0,49
8	51	113,6	+ 25	+ 23,5	625	552,25	587,5	+ 1,5	2,25
9	68	125,1	+ 42	+ 35,0	1764	1225,00	1470,0	+ 7,0	49,00
Σ	234	811,3	− 80 / + 80	− 70,2 / + 70,6	4060	3084,00	3534,8	− 9,8 / + 9,4	74,40
			0	+ 0,4				− 0,4	

TABLE 13.

$X_{1(j_1)}$	$\overline{X}_{(j_1)|1}$	$\tilde{X}_{(j_1)|1}$
0	66,7	67,5
4	71,0	71,0
10	76,3	76,2
15	80,6	80,6
21	85,7	85,8
29	92,9	92,7
36	99,4	98,8
51	113,6	111,9
68	125,1	126,7
Σ	(811,3)	(811,2)

corroborates our assumption that the dependence of sodium nitrate solubility on temperature is a linear one. The correlation equation expressing this dependence has the form

$$\tilde{X}_{(j_1)|1} = 90.1 + 0.999 \cdot \frac{18,512}{21.239} (X_{1(j_1)} - 26) =$$
$$= 67,5 + 0.87\, X_{1(j_1)}$$

with standard error

$$\sigma_{2.1} = 18.512 \sqrt{1 - 0.999^2} = 0.83.$$

This equation coincides with the one derived by D. I. Mendeleev.

The probable values, $\tilde{X}_{(j_1)|1}$, of sodium nitrate solubility, given by this equation, for given values, $X_{1(j_1)}$, of temperature, are shown in Table 13. In comparing them with the observed values, $\overline{X}_{(j_1)|1}$, we see virtually perfect coincidence between them; the straight line graphing the correlation equation passes very close to the observed points (Fig. 4).

2. Computing Second-Order Correlation Equations

We now consider the results of dynamic flattening of copper cylinders ($h_0 = 13$ mm, $d_0 = 8$ mm) by the impact of a falling steel block weighing 3.820 kg (from data of N. A. Shaposhnikov). The results are given in Table 14, where X_1 is the absolute residual flattening of the cylinder (mm) and X_2 is the total energy of the steel block absorbed by the cylinder (kgm).

The data of Table 14 are graphed in Fig. 4. It is clear from this figure that the relationship between the variables is not a linear one.

In order to set up the correlation equation, we find the necessary moments [columns (4)–(11) of Table 14].

In the present case, the mean values are

$$\overline{X}_1 = 4,33 \text{ mm}, \quad \overline{X}_2 = 9,653 \text{ kgm}.$$

After finding the deviations, x_1 and x_2, from these mean values, we form the columns headed x_1^2, x_2^2, $x_1 x_2$, x_1^3, x_1^4, $x_1^2 x_2$. The sums of these columns, divided by the number in the sample, give the corresponding central moments, from which we then find the standard moments (cf., the bottom row of Table 14).

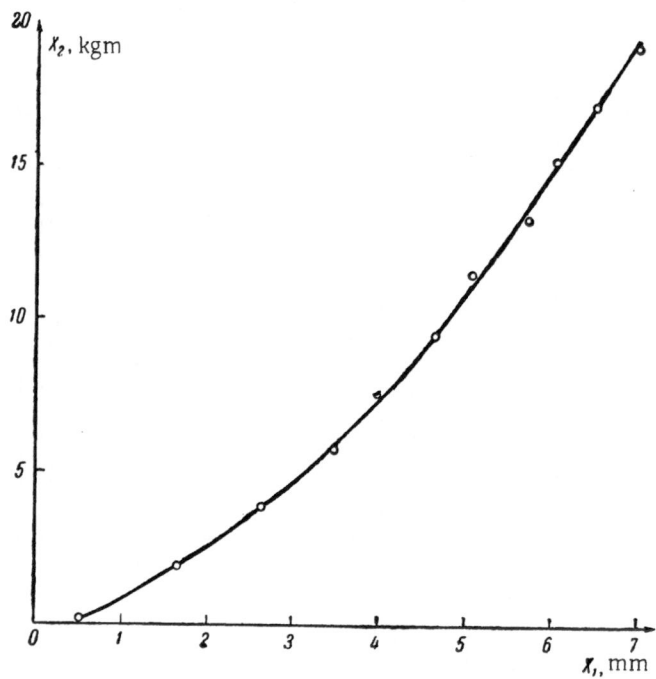

Fig. 5. Dependence of total energy of the steel block absorbed by a copper cylinder, X_2, on the absolute residual flattening of the cylinder, X_1.

TABLE 14. Layout of the Computation of Moments for Nonuniformly Spaced Variable Values

№	X_1	X_2	x_1	x_2	x_1^2	x_2^2	$x_1 x_2$	x_1^3	x_1^4	$x_1^2 x_2$
(1)	(2)	(3)	(4)	(5)	(6)	(7)	(8)	(9)	(10)	(11)
1	0,54	0,957	− 3,79	− 8,696	14,3641	75,6204	+ 32,9578	− 54,4399	206,3274	− 124,9102
2	1,68	1,916	− 2,65	− 7,737	7,0225	59,8612	+ 20,5030	− 18,6096	49,3155	− 54,3331
3	2,60	3,830	− 1,73	− 5,823	2,9929	33,9073	+ 10,0738	− 5,1777	8,9575	− 17,4277
4	3,42	5,743	− 0,91	− 3,910	0,8281	15,2881	+ 3,5581	− 0,7536	0,6857	− 3,2379
5	4,04	7,655	− 0,29	− 1,998	0,0841	3,9920	+ 0,5794	− 0,0244	0,0071	− 0,1680
6	4,64	9,568	+ 0,31	− 0,085	0,0961	0,0072	− 0,0264	+ 0,0298	0,0092	− 0,0082
7	5,22	11,480	+ 0,89	+ 1,827	0,7921	3,3379	+ 1,6260	+ 0,7050	0,6274	+ 1,4472
8	5,72	13,392	+ 1,39	+ 3,739	1,9321	13,9801	+ 5,1972	+ 2,6856	3,7330	+ 7,2241
9	6,18	15,304	+ 1,85	+ 5,651	3,4225	31,9338	+ 10,4544	+ 6,3316	11,7135	+ 19,3405
10	6,59	17,215	+ 2,26	+ 7,562	5,1076	57,1838	+ 17,0901	+ 11,5432	26,0876	+ 38,6237
11	6,99	19,127	+ 2,66	+ 9,474	7,0756	89,7567	+ 25,2008	+ 18,8211	50,0641	+ 67,0342
Σ	47,62	106,187	− 9,37 / + 9,36	− 28,249 / + 28,253	43,7177	384,8685	− 0,0264 / + 127,2406	− 79,0052 / + 40,1163	357,5280	− 200,0851 / + 133,6697
	$\overline{X}_1 = 4,33$	$\overline{X}_2 = 9,653$	− 0,01	+ 0,004	$\mu_{210} = 3,9743$	$\mu_{012} = 34,9880$	$\mu_{111} = 11,5649$	$\mu_{310} = -3,5353$	$\mu_{410} = 32,5025$	$\mu_{211} = -6,0378$
	—	—	$m_{110} = -0,0010$	$m_{011} = +0,0004$	$\sigma_1 = 1,9936$	$\sigma_2 = 5,9151$	$r_{111} = +0,9807$	$r_{310} = -0,4462$	$r_{410} = 2,0577$	$r_{211} = -0,2568$

18

TABLE 15. The Quantities a_1 and b_1 and the Criteria
for the Correlation Equation (from the Totals of Table 14)

$$a_1 = 2.0577 - (-0.4462)^2 - 1 = 0.8590,$$
$$b_1 = -0.2568 - 0.9807 \cdot (-0.4462) = +0.1808;$$
$$\zeta_1 = 1 - (0.9807)^2 = 0.0382,$$

$$\sigma_{\zeta_1} = 0.0618, \quad \frac{\zeta_1}{\sigma_{\zeta_1}} = 0.62;$$

$$\zeta_2 = 0.0382 - \frac{(0.1808)^2}{0.8590} = 0.0001,$$

$$\sigma_{\zeta_2} = 0.0033, \quad \frac{\zeta_2}{\sigma_{\zeta_2}} = 0.03.$$

From these moments we find the quantities a_1 and b_1, as well as the criteria for the correlation equations (Table 15), and we immediately conclude that a second-order correlation equation will suffice for expressing the relationship between the variables under study. This equation has the form

$$r^{(2)}_{(j_1)|1} = 0.9807\ \xi_{1(j_1)} + \frac{0.1808}{0.8590}(\xi^2_{1(j_1)} + 0.4462\,\xi_{1(j_1)} - 1) =$$
$$= -0.2105 + 1.0746\,\xi_{1(j_1)} + 0.2105\,\xi^2_{1(j_1)}$$

or

$$\tilde{X}_{(j_1)|1} = 9.653 + 5.9151\left\{ -0.2105 + 1.0746 \cdot \frac{X_{1(j_1)} - 4.33}{1.9936} + \right.$$
$$\left. + 0.2105 \cdot \frac{X^2_{1(j_1)} - 8.66\,X_{1(j_1)} + 18.7489}{3.9743} \right\} = 0.4805 + 0.4732\,X_{1(j_1)} +$$
$$+ 0.3135\,X^2_{1(j_1)}.$$

The probable values of the steel ball's total energy, as computed by this equation, are shown in the final column of Table 16.

3. Computing Third-Order Correlation Equations

Now, we set up the correlation equation which expresses the variation of compression, X_2, of an alcohol—water solution as a function of the percent alcohol content, X_1, at 0°C (Table 17). The tabular data were obtained and investigated by D. I. Mendeleev. By graphing these data (Fig. 6), we see that there is a nonlinear relationship between solution compression and percentage alcohol content.

TABLE 16

№	$X_{1(j_1)}$	$\bar{X}_{(j_1)|1}$	$\tilde{X}_{(j_1)|1}$
1	0,54	0,957	0,827
2	1,68	1,916	2,160
3	2,60	3,830	3,830
4	3,42	5,743	5,766
5	4,04	7,655	7,509
6	4,64	9,568	9,426
7	5,22	11,480	11,493
8	5,72	13,392	13,444
9	6,18	15,304	15,378
10	6,59	17,215	17,214
11	6,99	19,127	19,106
Σ	—	(106,187)	(106,153)

TABLE 17

№	X_1, %	X_2
(1)	(2)	(3)
1	39,9	4,0638
2	40,1	4,0692
3	42,0	4,1138
4	43,8	4,1330
5	45,0	4,1459
6	45,7	4,1495
7	46,2	4,1478
8	47,9	4,1349
9	49,5	4,1189
10	50,3	4,1092
11	51,8	4,0814
12	53,9	4,0285
Σ	556,1	49,2959

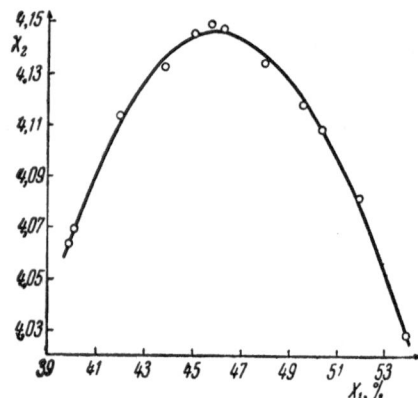

Fig. 6. Variation of compression, X_2, of an alcohol—water solution as a function of the percent content, X_1, of alcohol in the solution.

In order to ascertain the order of the correlation equation expressing the relationship in question, we again use Chebyshev's method.

We first find the means of the observed values of the quantities in question. By dividing the totals of columns (2) and (3) of Table 17 by the number of observations, we find

$$\overline{X}_1 = \frac{\Sigma X_1}{n} = 46.34167,$$

$$X_2 = \frac{\Sigma X_2}{n} = 4.10799.$$

Based on these mean values we determine the initial values, $X_{1(a)}$ and $X_{2(a)}$, relative to which we shall compute the moments necessary for setting up the correlation equations. The closer the initial values are chosen to the corresponding mean values, the easier will be all the following calculations. Thus, it is necessary so to choose $X_{1(a)}$ that the deviation

$$x_{1(j_1)} = X_{1(j_1)} - X_{1(a)}$$

is limited to two places. Then, in order to find the powers of these deviations, one may use published tables of powers of integers (cf., the Appendix, Table I. Also, Mitropolsky, "Short Mathematical Tables," Third Edition, Moscow, "Nauka," 1965). As for the other deviations

$$x_{2(j_2)} = X_{2(j_2)} - X_{2(a)},$$

they are to be chosen with the same accuracy as the observed data.

TABLE 18. Layout of the Computation of Moments for Nonuniformly Spaced Variable Values

№	x_1	x_2	x_1^2	x_2^2	$x_1 x_2$	x_1^3	x_1^4	$x_1^3 x_2$	x_1^5	x_1^6	$x_1^4 x_2$
(1)	(2)	(3)	(4)	(5)	(6)	(7)	(8)	(9)	(10)	(11)	(12)
1	− 6,4	− 0,0442	40,96	0,001954	+ 0,28288	− 262,144	1677,72	− 1,81043	− 10737,4	68719	+ 11,58676
2	− 6,2	− 0,0388	38,44	0,001505	+ 0,24056	− 238,328	1477,63	− 1,49147	− 9161,3	56800	+ 9,24713
3	− 4,3	+ 0,0058	18,49	0,000034	− 0,02494	− 79,507	341,88	+ 0,10724	− 1470,1	6321	− 0,46114
4	− 2,5	+ 0,0250	6,25	0,000625	− 0,06250	− 15,625	39,06	+ 0,15625	− 97,7	244	− 0,39062
5	− 1,3	+ 0,0379	1,69	0,001436	− 0,04927	− 2,197	2,86	+ 0,06405	− 3,7	5	− 0,08327
6	− 0,6	+ 0,0415	0,36	0,001722	− 0,02490	− 0,216	0,13	+ 0,01494	− 0,1	0	− 0,00896
7	− 0,1	+ 0,0398	0,01	0,001584	− 0,00398	− 0,001	0,00	+ 0,00040	0,0	0	− 0,00004
8	+ 1,6	+ 0,0269	2,56	0,000724	+ 0,04304	4,096	6,55	+ 0,06886	10,5	17	+ 0,11018
9	+ 3,2	+ 0,0109	10,24	0,000119	+ 0,03488	32,768	104,86	+ 0,11162	335,5	1074	+ 0,35717
10	+ 4,0	+ 0,0012	16,00	0,000001	+ 0,00480	64,000	256,00	+ 0,01920	1024,0	4096	+ 0,07680
11	+ 5,5	− 0,0266	30,25	0,000708	− 0,14630	166,375	915,06	− 0,80465	5032,8	27681	− 4,42558
12	+ 7,6	− 0,0795	57,76	0,006320	− 0,60420	438,976	3336,22	− 4,59192	25355,3	192700	− 34,89859
	− 21,4	− 0,1891	—	—	− 0,91609	− 598,018	—	− 8,69847	− 21470,3	—	− 40,26820
	+ 21,9	+ 0,1890	—	—	+ 0,60616	+ 706,215	—	+ 0,54256	+ 31758,1	—	+ 21,37804
Σ	+ 0,5	− 0,0001	223,01	0,016732	− 0,30993	+ 108,197	8157,97	− 8,15591	+ 10287,8	357657	− 18,89016

The computations of the moments are set out in Table 18.

To set up the first-order correlation equation, we fill in the first six columns of the layout. From the totals of these columns, we find

$$m_{1|0} = \frac{\Sigma x_1}{n} = +\,0.04167, \qquad m_{0|1} = \frac{\Sigma x_2}{n} = -\,0.00001,$$

$$m_{2|0} = \frac{\Sigma x_1^2}{n} = 18.584167, \qquad m_{0|2} = \frac{\Sigma x_2^2}{n} = 0.001394,$$

$$m_{1|1} = \frac{\Sigma x_1 x_2}{n} = -\,0.02583.$$

Whence,

$$\mu_{2|0} = m_{2|0} - m_{1|0}^2 = 18.582430, \qquad \mu_{0|2} = m_{0|2} - m_{0|1}^2 = 0.001394,$$
$$\sigma_1 = \sqrt{\mu_{2|0}} = 4.3107, \qquad \sigma_2 = \sqrt{\mu_{0|2}} = 0.0373,$$
$$r_{1|1} = \frac{m_{1|1} - m_{1|0}m_{0|1}}{\sigma_1 \sigma_2} = -\,0.1606.$$

By substituting the values found for $\overline{X}_1$, $\overline{X}_2$, σ_1, σ_2, and $r_{1|1}$ in (23), we obtain the first-order correlation equation for expressing the relationship between compression of the solution on the percentage alcohol content:

$$\tilde{X}_{(j_1)|1} = 4.17240 - 0.00139 X_{1\,(j_1)}.$$

For estimating the linearity of the functional relationship, we compute the linearity criterion (21). We get

$$\zeta_1 = 1 - (-\,0.1606)^2 = 0.9742$$

with standard error

$$\sigma_{\zeta_1} = \sqrt{\frac{0.9742}{12}} = 0.2850.$$

Since

$$\frac{\zeta_1}{\sigma_{\zeta_1}} = 3.418,$$

we cannot acknowledge that the relationship between our variables is linear. Consequently, it is necessary to turn to equations of higher order.

For setting up the second-order correlation equation, we need, in addition to the moments already calculated, the moments $r_{3|0}$, $r_{4|0}$, and $r_{2|1}$. For these, we fill in columns (7)–(9) of our layout. The totals of these columns give us

$$m_{3|0} = \frac{\Sigma x_1^3}{n} = +\,9.0164, \qquad m_{4|0} = \frac{\Sigma x_1^4}{n} = 679.8308,$$

$$m_{2|1} = \frac{\Sigma x_1^2 x_2}{n} = -\,0.6797$$

whence

$$\mu_{3|0} = 6.6917, \qquad \mu_{4|0} = 672.52, \qquad \mu_{2|1} = -\,0.6773,$$
$$r_{3|0} = 0.0835, \qquad r_{4|0} = 1.9650, \qquad r_{2|1} = -\,0.9772,$$
$$a_1 = 0.9580, \qquad b_1 = -\,0.9638.$$

By substituting these statistics in (25), we obtain the second-order correlation equation:

$$r_{(j_1)|1}^{(2)} = -0,1606\xi_{1(j_1)} + \frac{-0,9638}{0,9580}(\xi_{1(j_1)}^2 - 0,0835\xi_{1(j_1)} - 1) =$$
$$= 1,0061 - 0,0766\xi_{1(j_1)} - 1,0061\xi_{1(j_1)}^2.$$

For estimating the usefulness of this correlation equation, we compute the quadratic criterion of (27). This gives us

$$\zeta_2 = 0,9742 - \frac{(-0,9638)^2}{0,9580} = 0,0046$$

with standard error

$$\sigma_{\zeta_2} = 0,0196,$$

with

$$\frac{\zeta_2}{\sigma_{\zeta_2}} = 0,235.$$

By comparing the magnitudes of ζ_1 and ζ_2, and their ratios to the corresponding standard errors, we see a sharp reduction in the magnitude of the criteria for the correlation equations. This indicates that the second-order correlation equation apparently much more accurately expresses the relationship of solution compression to percentage alcohol content.

Desiring to express even more accurately the functional dependence in question, we turn to the setting up of the third-order correlation equation.

For this we must supplement the moments already computed by the moments $r_{5|0}$, $r_{6|0}$, and $r_{3|1}$. By filling out columns (10)-(12) of our layout, we find

$$m_5 = \frac{\Sigma x_1^5}{n} = 857,3167, \quad m_{6|0} = \frac{\Sigma x_1^6}{n} = 29804,75,$$
$$m_{3|1} = \frac{\Sigma x_1^3 x_2}{n} = -1,5742,$$

from whence

$$\mu_{5|0} = 715,69, \quad \mu_{6|0} = 29608, \quad \mu_{3|1} = -1,4892,$$
$$r_{5|0} = 0,4808, \quad r_{6|0} = 4,6145, \quad r_{3|1} = -0,4984,$$
$$a_2 = 0,2332, \quad a_3 = 0,7463, \quad b_2 = -0,1828,$$
$$\begin{vmatrix} a_1 & a_2 \\ a_2 & a_3 \end{vmatrix} = 0,6606. \quad \begin{vmatrix} a_1 & b_1 \\ a_2 & b_2 \end{vmatrix} = 0,0497,$$

and the third-order correlation equation of (29) will here take the form:

$$r_{(j_1)|1}^{(3)} = 1,0061 - 0,0766\xi_{1(j_1)} - 1,0061\xi_{1(j_1)}^2 +$$
$$+ \frac{0,0497}{0,6606}\left\{\xi_{1(j_1)}^3 - 1,9650\xi_{1(j_1)} - 0,0835 - \frac{0,2332}{0,9580}(\xi_{1(j_1)}^2 - 0,0835\xi_{1(j_1)} - 1)\right\}$$

or

$$r_{(j_1)|1}^{(3)} = 1,0181 - 0,2229\xi_{1(j_1)} - 1,0244\xi_{1(j_1)}^2 + 0,0752\xi_{1(j_1)}^3. \tag{39}$$

We now determine the cubic criterion of (31) in order to estimate the third-order correlation equation we have obtained. We have

$$\zeta_3 = 0,0046 - \frac{0,0025}{0,9580 \cdot 0,6606} = 0,0046 - 0,0039 = 0,0007$$

TABLE 19. Layout of the Computations for Smoothing Conditional Standard Moments Using Correlation Equation (39)

$X_{1(j_1)}$	$x_{1(j_1)} - m_{1\|0}$	$\xi_{1(j_1)}$	$\xi_{1(j_1)}^2$	$\xi_{1(j_1)}^3$	$1{,}0181 - 0{,}2229\xi_{1(j_1)} - 1{,}0244\xi_{1(j_1)}^2 + 0{,}0752\xi_{1(j_1)}^3$	$r_{(j_1)\|1}^{(3)}$
39,9	— 6,4417	— 1,4943	2,2320	— 3,3347	$1{,}0181 + 0{,}3331 - 2{,}2865 - 0{,}2508$	— 1,1861
40,1	— 6,2417	— 1,4479	2,0967	— 3,0360	$1{,}0181 + 0{,}3227 - 2{,}1479 - 0{,}2283$	— 1,0354
42,0	— 4,3417	— 1,0072	1,0141	— 1,0212	$1{,}0181 + 0{,}2245 - 1{,}0388 - 0{,}0768$	+ 0,1270
43,8	— 2,5417	— 0,5896	0,3476	— 0,2050	$1{,}0181 + 0{,}1314 - 0{,}3561 - 0{,}0154$	+ 0,7780
45,0	— 1,3417	— 0,3112	0,0968	— 0,0301	$1{,}0181 + 0{,}0694 - 0{,}0992 - 0{,}0023$	+ 0,9860
45,7	— 0,6417	— 0,1489	0,0222	— 0,0033	$1{,}0181 + 0{,}0332 - 0{,}0227 - 0{,}0002$	+ 1,0284
46,2	— 0,1417	— 0,0329	0,0011	— 0,0000	$1{,}0181 + 0{,}0073 - 0{,}0011 - 0{,}0000$	+ 1,0243
47,9	+ 1,5583	+ 0,3615	0,1307	+ 0,0472	$1{,}0181 - 0{,}0806 - 0{,}1339 + 0{,}0035$	+ 0,8071
49,5	+ 3,1583	+ 0,7327	0,5368	+ 0,3933	$1{,}0181 - 0{,}1633 - 0{,}5499 + 0{,}0296$	+ 0,3345
50,3	+ 3,9583	+ 0,9183	0,8433	+ 0,7744	$1{,}0181 - 0{,}2047 - 0{,}8639 + 0{,}0582$	+ 0,0077
51,8	+ 5,4583	+ 1,2662	1,6028	+ 2,0291	$1{,}0181 - 0{,}2822 - 1{,}6419 + 0{,}1526$	— 0,7534
53,9	+ 7,5583	+ 1,7534	3,0730	+ 5,3870	$1{,}0181 - 0{,}3908 - 3{,}1480 + 0{,}4051$	— 2,1156

TABLE 20

$X_{1(j_1)}$	$\overline{X}_{(j_1)\|1}$	$\tilde{X}_{(j_1)\|1}$
39,9	4,0638	4,0638
40,1	4,0692	4,0694
42,0	4,1138	4,1127
43,8	4,1330	4,1370
45,0	4,1459	4,1448
45,7	4,1495	4,1464
46,2	4,1478	4,1462
47,9	4,1349	4,1381
49,5	4,1189	4,1205
50,3	4,1092	4,1083
51,8	4,0814	4,0799
53,9	4,0285	4,0291
Σ	(49,2959)	(49,2962)

with standard error

$$\sigma_{\zeta_3} = 0{,}0076,$$

with

$$\frac{\zeta_3}{\sigma_{\zeta_3}} = 0.092.$$

In view of the insignificant size of ζ_3, we can halt at our third-order correlation equation in (39). The computations for smoothing the conditional standard moments by this equation are given in Table 19.

To determine the probable values, $\tilde{X}_{(j_1)\|1}$, of solution compression as a function of percentage alcohol content, $X_{1(j_1)}$, we have, in accordance with (14), the equation

$$\tilde{X}_{(j_1)\|1} = 4.1080 + r_{(j_1)\|1}^{(3)} \cdot 0.0373.$$

The results of these computations are given in Table 20. By plotting the tabular data on our graph (Fig. 6), we see that there is practically complete coincidence between the experimental values and the points computed by the correlation equation.

The derivation of correlation equation (39) makes it possible to solve the important problem of maximal compression of the solution.

For this, it is necessary to take the first derivative of the function

$$r_{(j_1)\|1}^{(3)} = k_0 + k_1 \xi_{1(j_1)} + k_2 \xi_{1(j_1)}^2 + k_3 \xi_{1(j_1)}^3,$$

equate this derivative to zero, and then solve the resulting equation. We have

$$\xi_{1(j_1)} = \frac{-k_2 \pm \sqrt{k_2^2 - 3k_1 k_3}}{3k_3},$$

with which the function attains its maximum for the value of the variable $\xi_{1(j_1)}$ corresponding to a minus sign before the radical.

Since, in the case at hand,

$$k_1 = -0{,}2229, \quad k_2 = -1{,}0244, \quad k_3 = 0{.}0752,$$

the value of the variable $\xi_{1(j_1)}$ for which the function $r_{(j_1)}^{(3)}|_1$ has its maximum equals

$$\xi_{1(j_1)} = \frac{+1{,}0244 - \sqrt{1{,}0244^2 + 3 \cdot 0{.}2229 \cdot 0{.}0752}}{3 \cdot 0{.}0752} =$$
$$= \frac{1{,}0244 - 1{.}0487}{0{.}2256} = -0{.}1077.$$

By substituting this value of $\xi_{1(j_1)}$ in our correlation equation (39), we find the maximum value

$$r_{(j_1)|1}^{(3)} = 1{,}0301,$$

whence

$$\tilde{X}_{(j_1)|1\,(max)} = 4{.}14642.$$

The percentage alcohol content of the solution for which, at 0°C, the greatest compression occurs, equals

$$X_{1(j_1)} = \overline{X}_1 + \xi_{1(j_1)}\sigma_1 = 46{,}342 - 0{,}1077 \cdot 4{,}3107 = 45{,}88,$$

the same value obtained by Mendeleev.

CHAPTER III

COMPUTING CORRELATION EQUATIONS BY MEANS OF CHEBYSHEV NUMBERS

1. Chebyshev Numbers

The computation of correlation equations from small samples becomes particularly easy in those cases in which the observed (grouped) values of one (independent) variable are equally spaced and have identical weights. In these cases, the correlation equations may be computed either by Chebyshev numbers or by the method of sums.

The computation of correlation equations of the required order by means of Chebyshev numbers consists in successively finding the terms of the following Chebyshev series:

$$_\lambda f(x) = \frac{\Sigma y_j}{n} + \frac{\Sigma y_j \psi_1(x_j)}{\Sigma \psi_1^2(x_j)} \psi_1(x) +$$
$$+ \frac{\Sigma y_j \psi_2(x_j)}{\Sigma \psi_2^2(x_j)} \psi_2(x) + \ldots + \frac{\Sigma y_j \psi_\lambda(x_j)}{\Sigma \psi_\lambda^2(x_j)} \psi_\lambda(x), \tag{40}$$

or

$$_\lambda f(x) = k_0 + k_1 \psi_1(x) + k_2 \psi_2(x) + \ldots + k_\lambda \psi_\lambda(x). \tag{41}$$

In this series, each of the coefficients, k_λ, for ψ_λ, is the sum of products of the values y_j by the values of the corresponding functions, $\psi_\lambda(x_j)$, divided by the sum of the squares of the values $\psi_\lambda(x_j)$, wherein the first term of the series is the average of the n values of y_j.

The functions $\psi_\lambda(x)$ satisfy the relationship

$$\psi_{\lambda+1}(x) = \psi_1(x) \psi_\lambda(x) - \frac{\lambda^2(n^2 - \lambda^2)}{4(2\lambda - 1)(2\lambda + 1)} \psi_{\lambda-1}(x), \tag{42}$$

with

$$\psi_0(x) = 1, \tag{43}$$

$$\psi_1(x) = x - \frac{n+1}{2}. \tag{44}$$

Based on these formulas, we find, in particular

$$\left. \begin{aligned}
\psi_2(x) &= \psi_1^2(x) - \frac{n^2 - 1}{12}, \\
\psi_3(x) &= \psi_1^3(x) - \frac{3n^2 - 7}{20} \psi_1(x), \\
\psi_4(x) &= \psi_1^4(x) - \frac{3n^2 - 13}{14} \psi_1^2(x) + \frac{3(n^2 - 1)(n^2 - 9)}{560}, \\
\psi_5(x) &= \psi_1^5(x) - \frac{5(n^2 - 7)}{18} \psi_1^3(x) + \frac{15n^4 - 230n^2 + 407}{1008} \psi_1(x).
\end{aligned} \right\} \tag{45}$$

The values of $\psi_\lambda(x_j)$, when $x_j = j$, $j = 1, \ldots, n$, for some given value of n, are called Chebyshev numbers. Values of Chebyshev numbers are given in Table II of the Appendix. Tabular values are provided for all values of n from 3 to 50. The Chebyshev numbers for n from 3 to 12 are given in their entirety; starting with n = 13, in order to conserve space, the values of $\psi_\lambda(x_j)$ are provided only for the second half of the series $x_1, ..., x_n$; the tabular values for the first half of any series are identical with those of the second half, for even λ, and have identical magnitudes, but reversed signs, for odd λ.

Computation of the values of the $\psi_\lambda(x_1)$, already premultiplied by their coefficients, is performed in accordance with the formulas

$$\left.\begin{aligned}
2\phi_1 &= -(n+1)+2x, \\
3\psi_2 &= \frac{3\cdot(2\psi_1)^2-(n^2-1)}{4}, \\
\frac{10}{3}\psi_3 &= \frac{5\cdot3\psi_2-(n^2-4)}{9}\cdot2\psi_1, \\
\frac{35}{12}\psi_4 &= \frac{20\cdot3\psi_2[7\cdot3\psi_2-(n^2-16)]-7(n^2-1)(n^2-4)}{432}, \\
\frac{21}{10}\psi_5 &= \frac{28\cdot3\psi_2[3\cdot3\psi_2-(n^2-16)]-(n^2+47)(n^2-4)}{720}\cdot2\psi_1.
\end{aligned}\right\} \tag{46}$$

A sum $\sum_j \psi_\lambda^2(x_j)$ is computed by the formula

$$\sum_j \psi_\lambda^2(x_j) = \frac{(\lambda!)^2 \, n\,(n^2-1)\,(n^2-4)\ldots(n^2-\lambda^2)}{[(2\lambda-1)!!]^2\,2^{2\lambda}\,(2\lambda+1)}. \tag{47}$$

[Trans. note: The American reader may not be familiar with the "double factorial" notation found in the denominator of equation (47). The ordinary factorial of n, n!, is the product of integer n by a l l smaller integers, down to unity; the so-called "skip factorial," n!!, is the product of integer n by e v e r y o t h e r smaller integer down to unity (for n odd), or down to 2 (for even n). Thus 5! is $5\cdot4\cdot3\cdot2\cdot1=120$; $5!!=5\cdot3\cdot1=15$, not to be confused with $(5!)!=(120)!$, a somewhat larger number. ESS]

In particular,

$$\left.\begin{aligned}
\sum_j \psi_1^2(x_j) &= \frac{n(n^2-1)}{2^2\cdot3}, \\
\sum_j \psi_2^2(x_j) &= \frac{n(n^2-1)(n^2-4)}{2^2\cdot3^2\cdot5}, \\
\sum_j \psi_3^2(x_j) &= \frac{n(n^2-1)(n^2-4)(n^2-9)}{2^4\cdot5^2\cdot7}, \\
\sum_j \psi_4^2(x_j) &= \frac{n(n^2-1)(n^2-4)(n^2-9)(n^2-16)}{2^2\cdot3^2\cdot5^2\cdot7^2}, \\
\sum_j \psi_5^2(x_j) &= \frac{n(n^2-1)(n^2-4)(n^2-9)(n^2-16)(n^2-25)}{2^4\cdot3^4\cdot7^2\cdot11}.
\end{aligned}\right\} \tag{48}$$

We note here that, for convenience in carrying out the computations of correlation equations, Table 2 in the Appendix gives the numbers $\psi_1, \psi_2, \psi_3, \psi_4,$ and ψ_5 premultiplied by the corresponding coefficients, $C_1,$ $C_2, C_3, C_4,$ and C_5. The price one pays for thus having exclusively integral entries in the table is that, to obtain some given term in the Chebyshev series, one must multiply the partial term

$$\frac{\Sigma\,C_\lambda\psi_\lambda y_j}{\Sigma\,(C_\lambda\psi_\lambda)^2}, \tag{49}$$

found by means of the tables, by

$$C_\lambda\psi_\lambda(x), \tag{50}$$

X	x	y	y^2	$C_1\psi_1$	$C_1\psi_1 y$	$C_2\psi_2$	$C_2\psi_2 y$	$C_3\psi_3$	$C_3\psi_3 y$	$C_4\psi_4$	$C_4\psi_4 y$
(1)	(2)	(3)	(4)	(5)	(6)	(7)	(8)	(9)	(10)	(11)	(12)
4,5	1	63,00	3969,0000	−4	−252,00	+28	+1764,00	−14	−882,00	+14	+ 882,00
5,5	2	59,00	3481,0000	−3	−177,00	+ 7	+ 413,00	+ 7	+413,00	−21	−1239,00
6,5	3	57,87	3348,9369	−2	−115,74	− 8	− 462,96	+13	+752,31	−11	− 636,57
7,5	4	56,59	3202,4281	−1	− 56,59	−17	− 962,03	+ 9	+509,31	+ 9	+ 509,31
8,5	5	55,84	3118,1056	0	0	−20	−1116,80	0	0	+18	+1005,12
9,5	6	55,45	3074,7025	+1	+ 55,45	−17	− 942,65	− 9	−499,05	+ 9	+ 490,05
10,5	7	55,18	3044,8324	+2	+110,36	− 8	− 441,44	−13	−717,34	−11	− 606,98
11,5	8	54,87	3010,7169	+3	+164,61	+ 7	+ 384,09	− 7	−384,09	−21	−1152,27
12,5	9	54,25	2943,0625	+4	+217,00	+28	+1519,00	+14	+759,50	+14	+ 759,50
—	—	512,05	29192,7849	60	− 53,91	2772	+ 154,21	990	− 48,36	2002	+ 11,16

i.e., by the product of two factors, one of which, C_λ, is found at the head of the λ'th column of Table 2 for the given value of n, the other of which is the function $\psi_\lambda(x)$, the form of which, for $\lambda = 1, 2, 3, 4,$ and 5, was given earlier (cf., (43)-(45)).

The standard error, indicating the degree to which a correlation equation of a given order approximates to the observed data, is computed from the formula

$$\sigma_\lambda = \sqrt{\frac{\Sigma_\lambda}{n - (\lambda + 1)}}, \tag{51}$$

in which the sum of the squares of the differences between the observed values, y_j, and the values defined by the correlation equation of the given order, is found by means of the formula

$$\Sigma_\lambda = \Sigma_{\lambda-1} - \frac{(\Sigma\, C_\lambda\psi_\lambda y_j)^2}{\Sigma\, (C_\lambda\psi_\lambda)^2} \tag{52}$$

or

$$\Sigma_\lambda = \Sigma_{\lambda-1} - \frac{k_\lambda^2 \cdot \Sigma\, (C_\lambda\psi_\lambda)^2}{C_\lambda^2}, \tag{53}$$

where

$$\Sigma_0 = \Sigma(y_j - \overline{y})^2 = \Sigma\, y_j^2 - \frac{(\Sigma\, y_j)^2}{n}. \tag{54}$$

2. Successive Computation of Correlation Equations

We shall now determine, using Chebyshev numbers, the correlation equation expressing the functional dependence of yield strength of axial steel, $y(\sigma_B,$ kg/mm^2)* on toughness, X (a_k, kgm/cm^2).

---

*Here, y is the corresponding conditional mean.

TABLE 22. Computation of the Zero'th Order Correlation Equation $_0f(x)$

1) $\Sigma y = 512{,}05$,

2) $\bar{y} = \dfrac{\Sigma y}{n} = \dfrac{512{,}05}{9} = 56{,}8944$,

3) $_0f(x) = 56{,}8944$,

4) $\Sigma y^2 = 29192{,}7849$,

5) $\Sigma_0 = \Sigma y^2 - \dfrac{(\Sigma y)^2}{n} =$

$= 29192{,}7849 - \dfrac{512{,}05^2}{9} =$

$= 29192{,}7849 - 29132{,}8002 =$

$= 59{,}9847$,

6) $\sigma_0 = \sqrt{\dfrac{\Sigma_0}{n-1}} = \sqrt{\dfrac{59{,}9847}{8}} =$

$= \sqrt{7{,}4981} = 2{,}738$.

The proper layout for computing correlation equations using Chebyshev numbers has the form shown on Table 21.

In column (1) we inscribe the values of the variable X; for the computation to follow, these values are replaced by ordinal numers, to be used as the values of the independent variable x, and these numbers are written in column (2); in column (3) we enter, and sum, the values of the dependent variable, y; in column (4) we enter the squares of the values of y; in columns (5), (7), (9), and (11), we copy from Table II of the Appendix, in the section devoted to the given number $n = 9$, the number $C_1\psi_1$, $C_2\psi_2$, $C_3\psi_3$, and $C_4\psi_4$, while in columns (6), (8), (10), and (12), we enter the products of the corresponding y_j values by the numbers in the columns on the left, and then sum the products in each column.

The first four columns of Table 21 are filled in to permit computation of the zero'th-order correlation equation. For each unit increase in the order of the correlation equation, two more columns of the table are added.

The computations are laid out in sequence, as shown in Tables 22 through 26. As is clear from these tables, computation of the correlation equation of any order requires that six steps be performed.

For example, to compute the zero'th-order correlation equation, it is necessary to do the following (cf., Table 22): 1) find Σy, the sum of the entries in column (3); 2) find the mean, $\bar{y} = \Sigma y/n$; 3) write the correlation equation $_0f(x)$; 4) compute Σy^2, the sum of the entries in column (4); 5) using formula (54), $\Sigma_0 = \Sigma y^2 - (\Sigma y)^2/n$, determine the sum of the squares of the differences between the observed values, y, and the values defined by the zero'th-order correlation equation; 6) compute the standard error $\sigma_0 = \sqrt{\Sigma_0}/(n-1)$.

TABLE 23. Computation of the First-Order Correlation Equation, $_1f(x)$

1) $\Sigma C_1\psi_1 y = -53{,}91$,

2) $\dfrac{\Sigma C_1\psi_1 y}{\Sigma (C_1\psi_1)^2} = \dfrac{-53{,}91}{60} = -0{,}8985$,

3) $\dfrac{\Sigma C_1\psi_1 y}{\Sigma (C_1\psi_1)^2} \cdot C_1 \cdot \left(x - \dfrac{n+1)}{2}\right) = -0{,}8985 \cdot 1 \cdot (x-5) =$

$= 4{,}4925 - 0{,}8985x$,

4) $\begin{array}{r} 56{,}8944 \\ + \ 4{,}4925 - 0{,}8985x \\ \hline 61{,}3869 - 0{,}8985x \end{array} = {_1f(x)}$,

5) $\Sigma_1 = \Sigma_0 - k_1^2 \cdot \dfrac{\Sigma (C_1\psi_1)^2}{C_1^2} = 59{,}9847 - (-0{,}8985)^2 \cdot \dfrac{60}{1} =$

$= 59{,}9847 - 48{,}4381 = 11{,}5466$,

6) $\sigma_1 = \sqrt{\dfrac{\Sigma_1}{n-2}} = \sqrt{\dfrac{11{,}5466}{7}} = \sqrt{1{,}6495} = 1{,}284$.

TABLE 24. Computation of the Second-Order Correlation Equation, $_2f(x)$

1) $\Sigma C_2 \psi_2 y = + 154{,}21$,

2) $\dfrac{\Sigma C_2 \psi_2 y}{\Sigma (C_2 \psi_2)^2} = \dfrac{+154{,}21}{2772} = + 0{,}0556$,

3) $\dfrac{\Sigma C_2 \psi_2 y}{\Sigma (C_2 \psi_2)^2} \cdot C_2 \cdot \left[\left(x - \dfrac{n+1}{2} \right)^2 - \dfrac{n^2-1}{12} \right] =$

$= + 0{,}0556 \cdot 3 \cdot \left[(x-5)^2 - \dfrac{80}{12} \right] =$

$= + 0{,}0556 \,(3x^2 - 30x + 55) = 3{,}0580 - 1{,}6680x + 0{,}1668x^2$,

4) $\begin{array}{l} 61{,}3869 - 0{,}8985x \\ \underline{3{,}0580 - 1{,}6680x + 0{,}1668x^2} \\ 64{,}4449 - 2{,}5665x + 0{,}1668x^2 = {}_2f(x), \end{array}$

5) $\Sigma_2 = \Sigma_1 - k_2^2 \cdot \dfrac{\Sigma (C_2 \psi_2)^2}{C_2^2} = 11{,}5466 - 0{,}1668^2 \cdot \dfrac{2772}{3^2} =$

$= 11{,}5466 - 8{,}5692 = 2{,}9774$,

6) $\sigma_2 = \sqrt{\dfrac{\Sigma_2}{n-3}} = \sqrt{\dfrac{2{,}9774}{6}} = \sqrt{0{,}4962} = 0{,}704$.

To compute the first-order correlation equation (cf., Table 23), one must: 1) fill in columns (5) and (6) of the layout, i.e., for the given number, n = 9, copy the numbers $C_1 \psi_1$ from Table II of the Appendix (into column (5) of the layout), multiply these by the corresponding values by y, putting the products in column (6), then sum column (6) to obtain $\Sigma C_1 \psi_1 y$; 2) divide the sum just found by the sum of the squares of the numbers $C_1 \psi_1$, this number being found at the foot of the first column of Table 2 in the Appendix devoted to n = 9, the result being $\Sigma C_1 \psi_1 y / \Sigma (C_1 \psi_1)^2$; 3) the factor just found is now multiplied by $C_1[x - (n+1)/2]$; 4) add the result just obtained to the right member of the zero'th-order correlation, thus obtaining the first-order correlation equation, $_1f(x)$; determine the sum Σ_1, for which it suffices to subtract from Σ_0 the square of the coefficient of x in the first-order correlation equation, k_1, multiplied by $\Sigma (C_1 \psi_1)^2 / C_1^2$; 6) calculate the first-order correlation equation's standard error, $\sigma_1 = \sqrt{\Sigma_1 / (n-2)}$.

TABLE 25. Computation of the Third-Order Correlation Equation, $_3f(x)$

1) $\Sigma C_3 \psi_3 y = - 48{,}36$,

2) $\dfrac{\Sigma C_3 \psi_3 y}{\Sigma (C_3 \psi_3)^2} = \dfrac{-48{,}36}{990} = - 0{,}0488$,

3) $\dfrac{\Sigma C_3 \psi_3 y}{\Sigma (C_3 \psi_3)^2} \cdot C_3 \cdot \left[\left(x - \dfrac{n+1}{2} \right)^3 - \dfrac{3n^2-7}{20} \left(x - \dfrac{n+1}{20} \right) \right] =$

$= - 0{,}0488 \cdot \dfrac{5}{6} \left[(x-5)^3 - \dfrac{236}{20} (x-5) \right] =$

$= 2{,}6840 - 2{,}5701x + 0{,}6100x^2 - 0{,}0407x^3$,

4) $\begin{array}{l} 64{,}4449 - 2{,}5665x + 0{,}1668x^2 \\ \underline{2{,}6840 - 2{,}5701x + 0{,}6100x^2 - 0{,}0407x^3} \\ 67{,}1289 - 5{,}1366x + 0{,}7768x^2 - 0{,}0407x^3 = {}_3f(x), \end{array}$

5) $\Sigma_3 = \Sigma_2 - k_3^2 \cdot \dfrac{\Sigma (C_3 \psi_3)^2}{C_3^2} = 2{,}9774 - (-0{,}0407)^2 \cdot \dfrac{990}{\left(\dfrac{5}{6} \right)^2} =$

$= 2{,}9774 - 2{,}3615 = 0{,}6159$,

6) $\sigma_3 = \sqrt{\dfrac{\Sigma_3}{n-4}} = \sqrt{\dfrac{0{,}6159}{5}} = \sqrt{0{,}1232} = 0{,}351$.

TABLE 26. Computation of the Fourth-Order Correlation Equation, $_4f(x)$

1) $\sum C_4 \psi_4 y = +11{,}16$,

2) $\dfrac{\sum C_4 \psi_4 y}{\sum (C_4 \psi_4)^2} = \dfrac{11{,}16}{2002} = +0{,}00557$,

3) $\dfrac{\sum C_4 \psi_4 y}{\sum (C_4 \psi_4)^2} \cdot C_4 \cdot \left[\left(x - \dfrac{n+1}{2} \right)^4 - \dfrac{3n^2 - 13}{14} \left(x - \dfrac{n+1}{2} \right)^2 + \right.$

$\left. + \dfrac{3(n^2 - 1)(n^2 - 9)}{560} \right] = 0{,}00557 \cdot \dfrac{7}{12} [(x-5)^4 -$

$- 16{,}4286 (x-5)^2 + 30{,}8571] = 0{,}79671 - 1{,}09107x +$

$+ 0{,}43411x^2 - 0{,}06500x^3 + 0{,}00325x^4$,

4) $_4f(x) = 67{,}9236 - 6{,}2277x + 1{,}2109x^2 - 0{,}1057x^3 + 0{,}00325x^4$,

5) $\Sigma_4 = \Sigma_3 - k_4^2 \cdot \dfrac{\sum (C_4 \psi_4)^2}{C_4^2} = 0{,}6159 - 0{,}00325^2 \cdot \dfrac{2002}{\left(\dfrac{7}{12}\right)^2} =$

$= 0{,}6159 - 0{,}0621 = 0{,}5538$,

6) $\sigma_4 = \sqrt{\dfrac{\Sigma_4}{n-5}} = \sqrt{\dfrac{0{,}5538}{4}} = 0{,}372$.

TABLE 27. Analysis of Variance of the Correlation Equation

Variance	Sum of Squares	Number of Degrees of freedom	Estimate of Variance
Total	59,9847	8	
Linear term	48,4381	1	48,4381
Square term	8,5692	1	8,5692
Cubic term	2,3615	1	2,3615
Higher term of the fourth-order equation	0,0621	1	0,0621
Residual	0,5538	4	0,1384

TABLE 28

$X_{1(J_1)}$	x	$\overline{X}_{(J_1)|1}$	$\tilde{X}_{(J_1)|1}$
4,5	1	63,00	62,73
5,5	2	59,00	59,64
6,5	3	57,87	57,61
7,5	4	56,59	56,41
8,5	5	55,84	55,78
9,5	6	55,45	55,48
10,5	7	55,18	55,28
11,5	8	54,87	54,91
12,5	9	54,25	54,15
Σ	—	(512,05)	(511,99)

In exactly the same fashion, we compute the correlation equations of third, fourth, and higher orders.

The question as to which order correlation equation one should stop at is resolved by comparing the standard errors of the correlation equations.

By comparing the standard errors of the correlation equations we have found (Tables 22-26), we see that these errors decrease sharply as one goes from the zero'th-order correlation equation to the first-order equation, then the second, finally the third. The transition to the fourth-order correlation equation is accompanied by no significant increase in accuracy. Consequently, it suffices to remain at the third-order correlation equation.

We arrive at the same conclusion by carrying out an analysis of variance of the correlation equation (cf., Table 27). To be sure, this is only a rough indication.

On the basis of Table 27, we find that the ratios of the estimates of variance of the linear, quadratic and cubic terms to the residual variance are significant; for example, for the cubic term we have

$$F = \frac{2.3615}{0.1384} = 17.06,$$

while the tabular value of the variance ratio at the 10% significance level, for $\nu_1 = 1$, $\nu_2 = 4$, is

$$F = 7.71.$$

[Trans. note to the reader: The author is here using the standard "F distribution," used, among other purposes, for testing the equality of two variances. Details may be found in any text on mathematical statistics. ESS]

We now find the variance ratio for the higher term of the fourth-order correlation equation.

Since, in computing F by the formula

$$F = \frac{s_1^2}{s_2^2}$$

it is assumed that

$$s_1^2 > s_2^2$$

we must, in our case, take $s_1^2 = 0.1384$ with $\nu_1 = 4$, and $s_2^2 = 0.0621$, with $\nu_2 = 1$. We have

$$F = \frac{0.1384}{0.0621} = 2.23,$$

while the tabular value of the variance ratio for the given numbers of degrees of freedom is

$$F = 224.6.$$

Consequently, the effect of the term in the fourth-order correlation equation is not significant.

Thus, to express the dependence of yield strength to rupture on the toughness of axial steel, we should content ourselves with the third-order correlation equation:

$$_3f(x) = 67.1289 - 5.1366x + 0.7768x^2 - 0.0407x^3.$$

The results of computations using this formula are provided in Table 28.

CHAPTER IV

COMPUTING CORRELATION EQUATIONS BY THE METHOD OF SUMS

1. The Method of Sums

In those cases when an investigation of the nature of the functional relationship at issue between random variables makes it possible to establish the order of the correlation equation to be computed, this computation is carried out particularly conveniently by the method of sums.

In computing correlation equations by the method of sums, the observed values of the first variable, X, are replaced by the deviations of these values from their mean value. These deviations are denoted by x, while the probable values of the second variable, y, are denoted by $\tilde{y}$.

In solving the normal equations for determining the coefficients of the correlation equation [cf., (5)], we find that, in the case of the first-order correlation equation

$$\tilde{y}^{(1)} = a_1 + b_1 x \tag{55}$$

the coefficients equal

$$\left.\begin{aligned} a_1 &= \frac{\Sigma y}{n}, \\ b_1 &= \frac{\Sigma xy}{\Sigma x^2}. \end{aligned}\right\} \tag{56}$$

In the case of the second-order correlation equation

$$\tilde{y}^{(2)} = a_2 + b_2 x + c_2 x^2 \tag{57}$$

the coefficients are

$$\left.\begin{aligned} a_2 &= \frac{1}{D_2}\begin{vmatrix} \Sigma y & \Sigma x^2 \\ \Sigma x^2 y & \Sigma x^4 \end{vmatrix}, \\ b_2 &= b_1, \\ c_2 &= \frac{1}{D_2}\begin{vmatrix} n & \Sigma y \\ \Sigma x^2 & \Sigma x^2 y \end{vmatrix}. \end{aligned}\right\} \tag{58}$$

In the case of the third-order equation

$$\tilde{y}^{(3)} = a_3 + b_3 x + c_3 x^2 + d_3 x^3 \tag{59}$$

the coefficients are

$$\left.\begin{aligned} a_3 &= a_2, \\ b_3 &= \frac{1}{D_3}\begin{vmatrix} \Sigma xy & \Sigma x^4 \\ \Sigma x^3 y & \Sigma x^6 \end{vmatrix}, \\ c_3 &= c_2, \\ d_3 &= \frac{1}{D_3}\begin{vmatrix} \Sigma x^2 & \Sigma xy \\ \Sigma x^4 & \Sigma x^3 y \end{vmatrix}. \end{aligned}\right\} \tag{60}$$

And, finally, for the fourth-order correlation equation

$$\tilde{y}^{(4)} = a_4 + b_4 x + c_4 x^2 + d_4 x^3 + e_4 x^4 \tag{61}$$

the coefficients equal

$$
\begin{aligned}
a_4 &= \frac{1}{D_4}
\begin{vmatrix}
\Sigma y & \Sigma x^2 & \Sigma x^4 \\
\Sigma x^2 y & \Sigma x^4 & \Sigma x^6 \\
\Sigma x^4 y & \Sigma x^6 & \Sigma x^8
\end{vmatrix}, \\
b_4 &= b_3, \\
c_4 &= \frac{1}{D_4}
\begin{vmatrix}
n & \Sigma y & \Sigma x^4 \\
\Sigma x^2 & \Sigma x^2 y & \Sigma x^6 \\
\Sigma x^4 & \Sigma x^4 y & \Sigma x^8
\end{vmatrix}, \\
d_4 &= d_3, \\
e_4 &= \frac{1}{D_4}
\begin{vmatrix}
n & \Sigma x^2 & \Sigma y \\
\Sigma x^2 & \Sigma x^4 & \Sigma x^2 y \\
\Sigma x^4 & \Sigma x^6 & \Sigma x^4 y
\end{vmatrix}.
\end{aligned}
\tag{62}
$$

The quantities in the denominators of these expressions are called the determinants of the distribution of the series of the first n integers, and equal

$$
\begin{aligned}
D_2 &=
\begin{vmatrix}
n & \Sigma x^2 \\
\Sigma x^2 & \Sigma x^4
\end{vmatrix}, \\
D_3 &=
\begin{vmatrix}
\Sigma x^2 & \Sigma x^4 \\
\Sigma x^4 & \Sigma x^6
\end{vmatrix}, \\
D_4 &=
\begin{vmatrix}
n & \Sigma x^2 & \Sigma x^4 \\
\Sigma x^2 & \Sigma x^4 & \Sigma x^6 \\
\Sigma x^4 & \Sigma x^6 & \Sigma x^8
\end{vmatrix}.
\end{aligned}
\tag{63}
$$

The sums of powers of the deviations from the mean

$$\Sigma x^2, \ \Sigma x^4, \ \Sigma x^6, \ \Sigma x^8$$

and the values of the determinants

$$D_2, \ D_3, \ D_4$$

were taken from Table 3 in the Appendix. To set up these tables, the sums of powers of the deviations of the integers from their mean were computed by the formulas

$$
\begin{aligned}
\Sigma x^2 &= \frac{n(n^2 - 1)}{12}, \\
\Sigma x^4 &= \Sigma x^2 \cdot \frac{3n^2 - 7}{20}, \\
\Sigma x^6 &= \Sigma x^2 \cdot \frac{3n^4 - 18n^2 + 31}{112}, \\
\Sigma x^8 &= \Sigma x^2 \cdot \frac{5n^6 - 55n^4 + 239n^2 - 381}{960}.
\end{aligned}
\tag{64}
$$

34

By means of these formulas, one can rewrite the determinants in (63) in the form

$$
\left.\begin{aligned}
D_2 &= \frac{n^2(n^2-1)(n^2-4)}{180}, \\
D_3 &= \frac{n^2(n^2-1)^2(n^2-4)(n^2-9)}{33\,600}, \\
D_4 &= \frac{n^2(n^2-1)^2(n^2-4)^2(n^2-9)(n^2-16)}{7\,938\,000}.
\end{aligned}\right\}
\tag{65}
$$

The computation of the quantities

$$
\Sigma y,\ \Sigma xy,\ \Sigma x^2 y,\ \Sigma x^3 y,\ \Sigma x^4 y
$$

is performed by the method of sums (cf., Table 49, Section 1). For this, it is necessary to differentiate between cases with odd numbers of values, and those with even numbers.

For an odd number of values, when the initial value coincides with the middle value (or class), these sums are easily obtained by the formula

$$
s_h = \sum_{j=1}^{k} \frac{x_j(x_j-1)\dots(x_j-h+1)}{h!}\, y_j.
\tag{66}
$$

We have

$$
\left.\begin{aligned}
\Sigma y &= s_0, \\
\Sigma xy &= d_1, \\
\Sigma x^2 y &= 2s_2 + s_1, \\
\Sigma x^3 y &= 6d_3 + 6d_2 + d_1, \\
\Sigma x^4 y &= 24 s_4 + 36 s_3 + 14 s_2 + s_1
\end{aligned}\right\}
\tag{67}
$$

[cf., equation (2), Appendix, Section 1].

When there is an even number of values, so that the initial value is chosen at a limit of a group, formula (66) takes the form

$$
s_h' = \sum_{j=1}^{k} \frac{\left(x_j+\frac{1}{2}\right)\left(x_j-\frac{1}{2}\right)\dots\left(x_j-h+\frac{3}{2}\right)}{h!}\, y_j.
\tag{68}
$$

From this we find

$$
\left.\begin{aligned}
\Sigma y &= s_0', \\
\Sigma xy &= d_1' - \frac{1}{2}d_0', \\
\Sigma x^2 y &= 2s_2' + \frac{1}{4}s_0', \\
\Sigma x^3 y &= 3(2d_3' + d_2') + \frac{1}{4}\left(d_1' - \frac{1}{2}d_0'\right), \\
\Sigma x^4 y &= 24(s_4' + s_3') + 5s_2' + \frac{1}{16}s_0'.
\end{aligned}\right\}
\tag{69}
$$

TABLE 29. Layout for Computing a Third-Order Correlation Equation
by the Method of Sums (odd number of values)

X	x	y	(1)	(2)	(3)	(4)	$x+1$	$(x+1)^3 y$
4,5	−4	63,00	63,00	63,00	63,00	63,00	−3	−1701,00
5,5	−3	59,00	122,00	185,00	248,00	—	−2	−472,00
6,5	−2	57,87	179,87	364,87	—	—	−1	−57,87
7,5	−1	56,59	236,46	—	—	—	0	0
8,5	0	55,84					+1	+55,84
9,5	+1	55,45	219,75	—	—	—	+2	+443,60
10,5	+2	55,18	164,30	327,67	—	—	+3	+1489,86
11,5	+3	54,87	109,12	163,37	217,62	—	+4	+3511,68
12,5	+4	54,25	54,25	54,25	54,25	54,25	+5	+6781,25
—	—	—	601,33	612,87	311,00	—	$\Sigma_3^{\bullet}$	+10051,36
—	—	—	547,42	545,29	271,87	—	—	—
—	s	512,05	1148,75	1158,16	—	—	—	—
—	d	—	−53,91	−67,58	−39,13	—	—	—

TABLE 30

$X_{1(J_1)}$	x	y	$\tilde{y}^{(3)}$
4,5	−4	63,00	62,73
5,5	−3	59,00	59,64
6,5	−2	57,87	57,61
7,5	−1	56,59	56,41
8,5	0	55,84	55,78
9,5	+1	55,45	55,49
10,5	+2	55,18	55,29
11,5	+3	54,87	54,93
12,5	+4	54,25	54,17
Σ	—	(512,05)	(512,05)

2. Computing Correlation Equations with an Odd Number of Values

We turn now to the computation of correlation equations by the method of sums, considering first the case of an odd number of values.

As an example, we set up the third-order correlation equation expressing the relationship of yield strength to rupture of axial steel, y (σ_B, kg/mm^2), on toughness, X (a_k, kgm/cm^2). The computations are laid out in Table 29.

From the table, we have

$$\Sigma y = s_0 = 512.05,$$
$$\Sigma xy = d_1 = -53.91,$$
$$\Sigma x^2 y = 2s_2 + s_1 = 3465.07,$$
$$\Sigma x^3 y = 6d_3 + 6d_2 + d_1 = -694.17.$$

The calculations are checked by use of the formula

$$\Sigma_3^{\bullet} = \Sigma y + 3\Sigma xy + 3\Sigma x^2 y + \Sigma x^3 y.$$

In our case,

$$\Sigma y = 512.05$$
$$3\Sigma xy = -161.73$$
$$3\Sigma x^2 y = 10395.21$$
$$\Sigma x^3 y = -694.17$$
$$\overline{\Sigma_3^{\bullet} = 10051.36.}$$

Consequently, the computations were performed correctly.

Using formulas (60), and extracting the values of the sums of powers of the deviations, x, and of the distribution determinants for n = 9, from Table 3 in the Appendix, we get

$$a_3 = \frac{512{,}05 \cdot 708 - 3465{,}07 \cdot 60}{2772} = 55{,}7818,$$

$$b_3 = \frac{-53{,}91 \cdot 9780 - (-694{,}17) \cdot 708}{85\,536} = -0{,}4182,$$

$$c_3 = \frac{9 \cdot 3465{,}07 - 60 \cdot 512{,}05}{2772} = 0{,}1669,$$

$$d_3 = \frac{60 \cdot (-694{,}17) - 708 \cdot (-53{,}91)}{85\,536} = -0{,}0407.$$

Substituting the values of the coefficients just found into equation (59), we get

$$\tilde{y}^{(3)} = 55{,}7818 - 0{,}4182x + 0{,}1669x^2 - 0{,}0407x^3.$$

By giving the deviation x the following values in this equation,

$$0, \quad \pm 1, \quad \pm 2, \quad \pm 3, \quad \pm 4,$$

we find the probable values of the yield strength for the corresponding values of axial steel toughness (cf., Table 30).

We had obtained the same values when we used Chebyshev numbers (see Table 28).

TABLE 31. Layout for Computing a Third-Order Correlation Equation
by the Method of Sums (even number of values)

X	x	y	(1)	(2)	(3)	(4)	$\|2(x+1)\|$	$[2(x+1)]^3 y$
0,04	$-\frac{7}{2}$	15,38	15,38	15,38	15,38	15,38	-5	$-1922{,}50$
0,08	$-\frac{5}{2}$	7,27	22,65	38,03	53,41	—	-3	$-196{,}29$
0,12	$-\frac{3}{2}$	4,38	27,03	65,06	—	—	-1	$-4{,}38$
0,16	$-\frac{1}{2}$	3,16	30,19	—	—	—	$+1$	$+3{,}16$
0,20	$+\frac{1}{2}$	2,93	20,47	—	—	—	$+3$	$+79{,}11$
0,24	$+\frac{3}{2}$	4,33	17,54	38,97	—	—	$+5$	$+541{,}25$
0,28	$+\frac{5}{2}$	4,99	13,21	21,43	29,65	—	$+7$	$+1711{,}57$
0,32	$+\frac{7}{2}$	8,22	8,22	8,22	8,22	8,22	$+9$	$+5992{,}38$
—	—	30,19	95,25	118,47	68,79	—	—	$-2123{,}17$
—	—	20,47	59,44	68,62	37,87	—	—	$+8327{,}47$
—	s'	50,66	154,69	187,09	—	—	Σ	6204,30
—	d'	$-9{,}72$	$-35{,}81$	$-49{,}85$	$-30{,}92$	—	$\Sigma_3^* = \dfrac{6204{,}30}{8} =$ $= 775{,}5375$	

TABLE 32

X	x	y	$\tilde{y}^{(3)}$
0,04	$-\dfrac{7}{2}$	15,38	14,95
0,08	$-\dfrac{5}{2}$	7,27	8,18
0,12	$-\dfrac{3}{2}$	4,38	4,28
0,16	$-\dfrac{1}{2}$	3,16	2,68
0,20	$+\dfrac{1}{2}$	2,93	2,79
0,24	$+\dfrac{3}{2}$	4,33	4,07
0,28	$+\dfrac{5}{2}$	4,99	5,92
0,32	$+\dfrac{7}{2}$	8,22	7,79
Σ	—	(50,66)	(50,66)

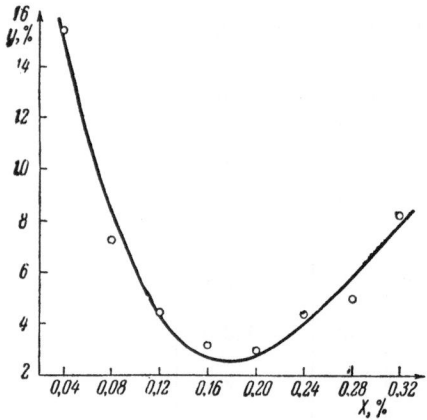

Fig. 7. Dependence of percentage reject of rails, y, on percentage of silicon content in steel, X. Note: With a decrease in the silicon content of the steel below 0.15 to 0.10%, the rails' brittleness increases due to inadequate reduction of the steel. With increasing silicon content above 0.25 to 0.30%, the percentage of rail rejects increases due to increasing rigidity of the steel from the alloying action of the silicon.

As an example of the computation of a correlation equation when the number of values is even, we shall set up the third-order correlation equation expressing the functional dependence of percentage reject of rails, y (based on drop-hammer testing), on the percentage of silicon content, X, of the steel (Table 31).

With an even number of values, the mean observed value of the first variable is found at the boundary between the two groups in the middle of the distribution. In our example, we have

$$\frac{0.04 + 0.32}{2} = 0.18,$$

which is found between groups 0.16 and 0.20. Therefore, the dividing line for the table of sums is drawn between these values, and the deviation of each value from the mean, expressed in units of groups (classes), will be

$$\pm\frac{1}{2}, \ \pm\frac{3}{2}, \ \pm\frac{5}{2}, \ \pm\frac{7}{2}.$$

Then Table 31 is set up exactly as was Table 29.

From Table 31, we have

$$\sum y = s_0' = 50.66,$$

$$\sum xy = d_1' - \frac{1}{2}d_0' = -35.81 - \frac{1}{2}\cdot(-9.72) = -30.95,$$

$$\sum x^2 y = 2s_2' + \frac{1}{4}s_0' = 2\cdot187.09 + \frac{1}{4}\cdot50.66 = 336.845,$$

$$\sum x^3 y = 3(2d_3' + d_2') + \frac{1}{4}\left(d_1' - \frac{1}{2}d_0'\right) =$$
$$= 3[2\cdot(-30.92) + (-49.85)] +$$
$$+ \frac{1}{4}\left| -35.81 - \frac{1}{2}\cdot(-9.72) \right| = -342.8075.$$

To verify these sums, we shift the initial value by one group in the direction of decreasing values and, for computational convenience, multiply by 2 the deviations, x + 1, thereby obtained. We may then apply the formula

$$\frac{1}{2^3}\sum[2(x+1)]^3 y = \sum y + 3\sum xy + 3\sum x^2 y + \sum x^3 y.$$

In our example,

$$\frac{1}{2^3}\sum[2(x+1)]^3 y = \frac{6204.30}{8} = 775.5375.$$

On the other hand,

$$\sum y = 50.66$$
$$3\sum xy = -92.85$$
$$3\sum x^2 y = 1160.535$$
$$\sum x^3 y = -342.8075$$
$$\overline{775.5375.}$$

Comparison of the results indicates that our computations were correct.

Employing now our formulas (60), and taking from Table III of the Appendix the values of the sums of powers of the deviations, x, and of the determinants of the distribution for n = 8, we find

$$a_3 = \frac{50.66 \cdot 388.5 - 336.845 \cdot 42}{1344} = \frac{3433.920}{1344} = 2.555,$$

$$b_3 = \frac{-30.95 \cdot 4187.625 - 342.8075 \cdot 388.5}{24\,948} = \frac{3573.720}{24\,948} = 0.1432,$$

$$c_3 = \frac{8 \cdot 386.845 - 42 \cdot 50.66}{1344} = \frac{967.040}{1344} = 0.7195,$$

$$d_3 = \frac{42 \cdot (-342.8075) + 388.5 \cdot 30.95}{24\,948} = \frac{-2363.840}{24\,948} = -0.09515.$$

Thus, for the given case, correlation equation (59) will have the form

$$\tilde{y}^{(3)} = 2.555 + 0.1432x + 0.7195x^2 - 0.09515x^3.$$

Giving to the deviations, x, the values

$$\pm \frac{1}{2}, \ \pm \frac{3}{2}, \ \pm \frac{5}{2}, \ \pm \frac{7}{2},$$

we obtain the probable values of the percent rejects of rails for the corresponding values of silicon content of steel (Table 32).

For the same example, the third-order correlation equation was also set up by means of Chebyshev numbers

$$_3f(x) = 25.15 - 12.110x + 2.0033x^2 - 0.09515x^3.$$

Giving to the deviations x the values here of

$$1, \ 2, \ \ldots \ 8,$$

we obtain the same probable values of percent rejects of rails.

The graph of the data in Table 32 on Fig. 7 indicates a good fit of the probable values of percent rail rejects to the experimental data.

CHAPTER V

COMPUTATION OF CORRELATION EQUATIONS OF A NONPOLYNOMIAL TYPE

1. Computing a Correlation Equation's Coefficients

Correlation equations in the form of polynomials of some order or another may be chosen to represent the relationship among random variables even in those cases when the character of the relationship under study is as yet unknown. By using Chebyshev's method, we may determine the order of the correlation equation providing the best fit to the graph of the observed relationship between the variables.

In those cases when a knowledge of the nature of the relationship between the random variables makes it possible to elucidate the character of this relationship, the type of correlation equation is selected in complete correspondence with this character. We shall now consider the method of computing the most important of these types of correlation equations.

The basic method of calculating the coefficients of a correlation equation is the method of least squares (cf., Chaper I, Section 1). For this, the correlation equation being sought is put into linear form

$$\tilde{y} = a + bx. \tag{70}$$

Then, from the observed data, one tries to obtain an approximate equality

$$y_j \approx a + bx_j \quad (j = \overline{1, n}), \tag{71}$$

by seeking coefficients a and b by the condition that the sum of the squares of the differences between the left and right members of these approximate equalities

$$\sum_{j=1}^{n} [y_j - (a + bx_j)]^2 \tag{72}$$

be minimized. *

By using the standard method of finding minimal values, we obtain the two normal equations

$$\left. \begin{array}{l} an + b\sum x = \sum y, \\ a\sum x + b\sum x^2 = \sum xy. \end{array} \right\} \tag{73}$$

Solving these equations, we find

$$a = \frac{A}{D}, \quad b = \frac{B}{D}. \tag{74}$$

---

*Expression (71) is used only in the case when the y_j all have identical weights.

TABLE 33. Computation of Determinants and Coefficients
of a Correlation Equation

№	x	y	x^2	xy	$\tilde{y}$
1	0	66,7	0	0	67,5
2	4	71,0	16	284,0	71,0
3	10	76,3	100	763,0	76,2
4	15	80,6	225	1209,0	80,6
5	21	85,7	441	1799,7	85,8
6	29	92,9	841	2694,1	92,7
7	36	99,4	1296	3578,4	98,8
8	51	113,6	2601	5793,6	111,9
9	68	125,1	4624	8506,8	126,7
Σ	234	811,3	10 144	24628,6	811,2

$$D = \begin{vmatrix} 9 & 234 \\ 234 & 10\,144 \end{vmatrix} = 9 \cdot 10\,144 - 234^2 = 36540,$$

$$A = \begin{vmatrix} 811,3 & 234 \\ 24628,6 & 10\,144 \end{vmatrix} = 811,3 \cdot 10144 - 24628,6 \cdot 234 = 2466734,8,$$

$$B = \begin{vmatrix} 9 & 811,3 \\ 234 & 24628,6 \end{vmatrix} = 9 \cdot 24628,6 - 234 \cdot 811,3 = 31813,2,$$

$$a = \frac{2466734,8}{36\,540} = 67,\tilde{5}, \qquad b = \frac{31813,2}{36\,540} = 0,87.$$

In the denominators of these expressions is the system's determinant

$$D = \begin{vmatrix} n & \sum x \\ \sum x & \sum x^2 \end{vmatrix}, \tag{75}$$

whose elements are the coefficients of a and b in the left members of the normal equations. The determinant

$$A = \begin{vmatrix} \sum y & \sum x \\ \sum xy & \sum x^2 \end{vmatrix} \tag{76}$$

is obtained from the system's determinant by replacing the elements of its first column, i.e., the coefficients of a, by the "constant" terms, the right members of the normal equations, and the determinant

$$B = \begin{vmatrix} n & \sum y \\ \sum x & \sum xy \end{vmatrix} \tag{77}$$

is obtained from the system's determinant by replacing by the constant terms the elements of the second column, i.e., the coefficients of b.

Let us, for example, set up correlation equation (70) for expressing the dependence of sodium nitrate solubility, y, on temperature changes, x, in degrees centigrade (Table 11).

The calculations of the determinants and the coefficients of the correlation are laid out in Table 33.

Thus, the correlation equation we sought takes the form

$$\tilde{y} = 67.5 + 0.87x$$

(cf., Chapter II, Section 1). The probable values of sodium nitrate solubility, computed from this equation, are given in the last column of Table 33.

TABLE 34. Layout for Calculating a Power-Type Correlation Equation

x	y	$\lg x$	$(\lg x)^2$	$\lg y$	$\lg x \cdot \lg y$	$\lg \tilde{y}$	$\tilde{y}$
(1)	(2)	(3)	(4)	(5)	(6)	(7)	(8)
1	56	0,00000	0,00000	1,74819	0,00000	1,74830	56,08
3	28	0,47712	0,22764	1,44716	0,69047	1,44445	27,83
5	20	0,69897	0,48856	1,30103	0,90938	1,30317	20,09
7	16	0,84510	0,71419	1,20412	1,01760	1,21011	16,22
9	14	0,95424	0,91057	1,14613	1,09368	1,14060	13,82
Σ	(134)	2,97543	2,34096	6,84663	3,71113	6,84663	(134,04)

2. The Power-Type Correlation Equation

Among the nonpolynomial types of correlation equations, the most frequently used is the power-type correlation equation

$$\tilde{y} = ax^b, \tag{78}$$

in which the independent variable is raised to a power, b, which, along with the multiplicative factor a, is a parameter of this equation.

By taking logarithms of both sides of this equation, we may bring it to a linear form

$$\lg \tilde{y} = \lg a + b \lg x. \tag{79}$$

The coefficients of equation (79) are found by the method of least squares. For this, we set up the two normal equations:

$$\left. \begin{array}{l} n \lg a + b \sum \lg x = \sum \lg y, \\ \lg a \sum \lg x + b \sum (\lg x)^2 = \sum \lg x \lg y. \end{array} \right\} \tag{80}$$

In this case, we have the three determinants:

$$\left. \begin{array}{l} D = \begin{vmatrix} n & \sum \lg x \\ \sum \lg x & \sum (\lg x)^2 \end{vmatrix}, \\ A = \begin{vmatrix} \sum \lg y & \sum \lg x \\ \sum \lg x \lg y & \sum (\lg x)^2 \end{vmatrix}, \\ B = \begin{vmatrix} n & \sum \lg y \\ \sum \lg x & \sum \lg x \lg y \end{vmatrix}. \end{array} \right\} \tag{81}$$

Based on these determinants, we find the coefficients of the correlation equation

$$\lg a = \frac{A}{D}, \quad b = \frac{B}{D}. \tag{82}$$

Let us consider the functional dependence of hygrometer readings, y (in scale divisions), on work-piece thickness, x (cm), for beech wood (Table 34).

A qualitative analysis of the phenomenon, and of the character of the curve fit by eye, make it reasonable to assume that the relationship in question must be expressed by equation (78). The computations are laid out in Table 34.

By substituting the sums thus found

$$n = 5, \quad \sum \lg x = 2.97543,$$
$$\sum (\lg x)^2 = 2.34096, \quad \sum \lg y = 6.84663,$$
$$\sum \lg x \lg y = 3.71113$$

in the determinants of (81), we get

$$D = \begin{vmatrix} 5 & 2.97543 \\ 2.97543 & 2.34098 \end{vmatrix} = 2.85162,$$

$$A = \begin{vmatrix} 6.84663 & 2.97543 \\ 3.71113 & 2.34098 \end{vmatrix} = 4.98548,$$

$$B = \begin{vmatrix} 5 & 6.84663 \\ 2.97543 & 3.71113 \end{vmatrix} = -1.81602.$$

The coefficients of (82) will thus equal

$$\lg a = \frac{4.98548}{2.85152} = 1.74830,$$
$$b = \frac{-1.81602}{2.85162} = -0.63684.$$

In that way, correlation equation (79) takes the form

$$\lg \tilde{y} = 1.74830 - 0.63684 \lg x.$$

By substituting in this equation the logarithms of the observed values, lg x, we obtain lg $\tilde{y}$, from which we then find $\tilde{y}$.

3. The Exponential-Type Correlation Equation

In those cases when the velocity of change of the quantity y, as a function of the variations in variable x, is generally proportional to the value of the quantity y itself, the dependence of y on x is best expressed by a correlation equation of the exponential type

$$\tilde{y} = ab^x. \tag{83}$$

Taking logarithms of both sides of this equation, we get

$$\lg \tilde{y} = \lg a + x \lg b. \tag{84}$$

The normal equations then have the form

$$\left. \begin{array}{l} n \lg a + \lg b \sum x = \sum \lg y, \\ \lg a \sum x + \lg b \sum x^2 = \Sigma x \lg y. \end{array} \right\} \tag{85}$$

TABLE 35. Layout for Computing an Exponential-Type Correlation Equation

X	x	y	x^2	$\lg y$	$x \lg y$	$\lg \tilde{y}$	$\tilde{y}$
1,65	1	122,7	1	2,08884	2,08884	2,06360	115,8
1,75	2	157,7	4	2,19783	4,39566	2,15588	143,2
1,85	3	181,2	9	2,25816	6,77448	2,24816	177,1
1,95	4	188,1	16	2,27439	9,09756	2,34044	219,0
2,05	5	284,3	25	2,45378	12,26890	2,43272	270,8
2,15	6	295,9	36	2,47114	14,82684	2,52500	335,0
2,25	7	415,7	49	2,61878	18,33146	2,61728	414,3
2,35	8	480,8	64	2,68196	21,45568	2,70956	512,4
2,45	9	603,8	81	2,78089	25,02801	2,80184	633,6
2,55	10	812,3	100	2,90972	29,09720	2,89412	783,7
2,65	11	1093,6	121	3,03886	33,42746	2,98640	969,2
2,75	12	1201,2	144	3,07962	36,95544	3,07868	1198,4
Σ	78	(5837,3)	650	30,85397	213,74753		(5772,5)

After having computed the determinants

$$D = \begin{vmatrix} n & \Sigma x \\ \Sigma x & \Sigma x^2 \end{vmatrix},$$

$$A = \begin{vmatrix} \Sigma \lg y & \Sigma x \\ \Sigma x \lg y & \Sigma x^2 \end{vmatrix}, \qquad (86)$$

$$B = \begin{vmatrix} n & \Sigma \lg y \\ \Sigma x & \Sigma x \lg y \end{vmatrix},$$

we find the coefficients of correlation equation (84):

$$\lg a = \frac{A}{D}, \qquad \lg b = \frac{B}{D}. \qquad (87)$$

The values of the sums of powers of integers, Σx and Σx^2, occurring in the determinants of (86), may be copied directly from Table IV in the Appendix. *

Let us set up an exponential-type correlation equation to express the dependence of yield strength under compression, y (σ_B, kg/cm^2), on weight-by-volume, X (γ, g/cm^3), of limestone (Table 35). For computational convenience, we replace the values of weight-by-volume by integers (second column of the table).

From the table we have

$$n = 12, \quad \Sigma x = 78, \quad \Sigma x^2 = 650,$$
$$\Sigma \lg y = 30.85397, \quad \Sigma x \lg y = 213.74753.$$

On the basis of these sums, we find the determinants

$$D = 12 \cdot 650 - 78^2 = 1716,$$
$$A = 30.85397 \cdot 650 - 213.74753 \cdot 78 = 3382.78876,$$
$$B = 12 \cdot 213.74753 - 78 \cdot 30.85397 = 158.35830.$$

---

*Here, $x_j = j$, $j = 1, \ldots, h$. It is assumed that one can arrange this by means of a linear transformation.

The correlation equation's coefficients will then equal

$$\lg a = \frac{3382,78876}{1716} = 1,97132,$$

$$\lg b = \frac{158,35830}{1716} = 0,09228.$$

Thus, the correlation equation has the form

$$\lg \tilde{y} = 1.97132 + 0.09228x.$$

By giving the variable x the values

$$1, \ 2, \ \ldots, \ 12,$$

we find $\lg \tilde{y}$; from these logarithms we then obtain the probable values, $\tilde{y}$. These computations are shown in the two last columns of Table 35.

To check our computations, we take, instead of x, the actual values of the first column: 1.65, 1.75, ..., 2.75. In this case, we obtain

$$\Sigma x = 26.40, \qquad \Sigma x^2 = 59,5100,$$
$$\Sigma \lg y = 30.85397, \quad \Sigma x \lg y = 69,19843,$$
$$D = 17,1600, \qquad A = 9,2812, \quad B = 15.2812,$$
$$\lg a = 0,54086, \qquad \lg b = 0.92287,$$

and the correlation equation takes the form

$$\lg \tilde{y} = 0.54086 + 0.92287x.$$

Giving, in this equation, the values 1.65, 1.75, ..., 2.75 to x, we obtain the same values for $\tilde{y}$ as in the first case.

A comparison of observed values of yield strength of limestone under compression with the probable values computed from our correlation equation shows a sufficiently good fit between them.

TABLE 36. Layout for Computing a Logarithmic-Type Correlation Equation

X	y	x	$\lg x$	$y \lg x$	$\tilde{y}$
(1)	(2)	(3)	(4)	(5)	(6)
20	9,3	1	0,00000	0,00000	7,94
30	14,1	2	0,30103	4,24452	15,07
40	18,3	3	0,47712	8,73130	19,25
50	21,8	4	0,60206	13,12491	22,21
60	24,2	5	0,69897	16,91507	24,50
70	26,3	6	0,77815	20,46534	26,38
80	28,1	7	0,84510	23,74731	27,97
90	29,6	8	0,90309	26,73146	29,34
100	30,9	9	0,95424	29,48602	30,55
110	32,0	10	1,00000	32,00000	31,64
120	32,9	11	1,04139	34,26173	32,62
130	33,7	12	1,07918	36,36837	33,57
140	34,4	13	1,11394	38,31954	34,34
150	35,0	14	1,14613	40,11455	35,10
160	35,6	15	1,17609	41,86880	35,81
Σ	(406,2)	—	12,11649	366,37892	(406,29)

4. The Logarithmic-Type Correlation Equation

A correlation equation of the logarithmic type has the form

$$\tilde{y} = a + b \lg x. \tag{88}$$

The term $b \lg x$ of this equation expresses the decreasing increase in the function as its argument increases, the characteristic feature of logarithmic curves, and the constant term, a, indicates the level from which the logarithmic curve (asymptotically) starts.

For the determination of the coefficients of equation (88), we set up the two normal equations:

$$\left. \begin{array}{l} an + b \sum \lg x = \sum y, \\ a \sum \lg x + b \sum (\lg x)^2 = \sum y \lg x. \end{array} \right\} \tag{89}$$

By computing the determinants

$$\left. \begin{array}{l} D = \begin{vmatrix} n & \sum \lg x \\ \sum \lg x & \sum (\lg x)^2 \end{vmatrix}, \\[12pt] A = \begin{vmatrix} \sum y & \sum \lg x \\ \sum y \lg x & \sum (\lg x)^2 \end{vmatrix}, \\[12pt] B = \begin{vmatrix} n & \sum y \\ \sum \lg x & \sum y \lg x \end{vmatrix}, \end{array} \right\} \tag{90}$$

we find

$$a = \frac{A}{D}, \qquad b = \frac{B}{D}. \tag{91}$$

Sometimes, in equation (88), a linear term, bx, is introduced, and the new logarithmic-type correlation equation takes the form

$$\tilde{y} = a + bx + c \lg x. \tag{92}$$

To determine the coefficients of this equation, we must set up three normal equations

$$\left. \begin{array}{l} na + b \sum x + c \sum \lg x = \sum y, \\ a \sum x + b \sum x^2 + c \sum x \lg x = \sum xy, \\ a \sum \lg x + b \sum x \lg x + c \sum (\lg x)^2 = \sum y \lg x. \end{array} \right\} \tag{93}$$

To solve these equations, it is necessary to calculate the determinants

$$\left. \begin{array}{l} D = \begin{vmatrix} n & \sum x & \sum \lg x \\ \sum x & \sum x^2 & \sum x \lg x \\ \sum \lg x & \sum x \lg x & \sum (\lg x)^2, \end{vmatrix}, \\[20pt] A = \begin{vmatrix} \sum y & \sum x & \sum \lg x \\ \sum xy & \sum x^2 & \sum x \lg x \\ \sum y \lg x & \sum x \lg x & \sum (\lg x)^2 \end{vmatrix}, \\[20pt] B = \begin{vmatrix} n & \sum y & \sum \lg x \\ \sum x & \sum xy & \sum x \lg x \\ \sum \lg x & \sum y \lg x & \sum (\lg x)^2 \end{vmatrix}, \\[20pt] C = \begin{vmatrix} n & \sum x & \sum y \\ \sum x & \sum x^2 & \sum xy \\ \sum \lg x & \sum x \lg x & \sum y \lg x \end{vmatrix} \end{array} \right\} \tag{94}$$

We find, based on these determinants,

$$a = \frac{A}{D}, \qquad b = \frac{B}{D}, \qquad c = \frac{C}{D}. \tag{95}$$

The values of the sums Σx and Σx^2 are transcribed from Table IV in the Appendix, and the values of the sums $\Sigma \lg x$, $\Sigma(x \lg x)$, and $\Sigma(\lg x)^2$, from Table V in the Appendix.

As an example, we shall compute correlation equation (88) for the series presenting the changes in mean height of oak trees, $y(m)$, as a function of growth, X (in years), in normal oak seed plantings. The computations are shown in Table 36.

On the basis of the sums thus obtained, we may, by using Table 5 in the Appendix, write the normal equations of (89) in the following form:

$$15a + 12.11649b = 406.2,$$
$$12.11649a + 11.40196b = 366.37892.$$

Then, we find

$$D = 24.22007, \quad A = 192.28963, \quad B = 573.96556.$$

Consequently,

$$a = 7.94, \quad b = 23.698.$$

Thus, the correlation equation being sought is

$$\tilde{y} = 7.94 + 23.698 \lg x.$$

By substituting the successive values of x in this equation (cf., columns (3) and (4) of the layout), we find the probable values, $\tilde{y}$, of oak height for the corresponding values of growth. These values are given in column (6) of the layout. In comparing the observed mean values, y, and the probable values, $\tilde{y}$, we see a satisfactorily complete coincidence of them.

5. The Periodic-Type Correlation Equation

A special type of correlation equation makes its appearance in investigations of certain periodic phenomena. Consider the following outline.

We suppose that, for n equally-spaced values of the variable x, equal to

$$0, 1 \cdot \frac{2\pi}{n}, \; 2 \cdot \frac{2\pi}{n}, \; \ldots, \quad h \cdot \frac{2\pi}{n}, \; \ldots, \quad (n-1) \cdot \frac{2\pi}{n}, \tag{96}$$

observations provided n values of a random variable, of equal weight

$$u_0, \quad u_1, \quad u_2, \; \ldots, \quad u_h, \; \ldots, \quad u_{n-1}. \tag{97}$$

We may then attempt to express the dependence of variable u on x by the equation

$$\tilde{u} = a_0 + \sum_{k=1}^{m} (a_k \cos kx + b_k \sin kx). \tag{98}$$

48

TABLE 37

Months	$t\,^{\circ}C$	Months	$t\,^{\circ}C$
1	−8,36	7	+17,42
2	−8,06	8	+15,79
3	−4,29	9	+10,76
4	+2,25	10	+4,81
5	+8,87	11	−1,29
6	+14,62	12	−6,03

In this equation, the number of unknown coefficients is 2m + 1. If

$$n > 2m + 1,$$

the coefficients a_k and b_k are found by the method of least squares, i.e., from the condition that

$$\sum_{h=0}^{n-1} (u_h - \tilde{u}_h)^2 \qquad (99)$$

have a minimal value.

To determine the coefficients a_k and b_k, we take into account that the functions

$$1, \quad \cos kx, \quad \sin kx \quad (k = \overline{1, \ m}),$$

where

$$m \leqslant \frac{n}{2},$$

comprise a system of orthogonal functions with respect to the system of equally-spaced values of x_h from the series of (96):

$$\sum_{h=0}^{n-1} \cos \frac{2hk\pi}{n} \cos \frac{2hl\pi}{n} \left\{ \begin{array}{ll} = 0, & \text{if} \quad k \neq l, \\ = \frac{n}{2}, & \text{if} \quad k = l \neq 0, \neq \frac{n}{2}, \end{array} \right. \qquad (100)$$

with similar equations in which one or both of the cosines is replaced by a sine.

To seek the coefficients of equation (98), we differentiate (99) with respect to a_k and b_k, and set the derivatives equal to zero. Thanks to the orthogonality properties expressed by (100), the solution of the normal equations turns out to be very easy. We have

$$\left.\begin{array}{l} a_0 = \dfrac{1}{n} \displaystyle\sum_{h=0}^{n-1} u_h, \\[2ex] a_k = \dfrac{2}{n} \displaystyle\sum_{h=0}^{n-1} u_h \cos kx_h, \\[2ex] b_k = \dfrac{2}{n} \displaystyle\sum_{h=0}^{n-1} u_h \sin kx_h. \end{array}\right\} \qquad (101)$$

The simplicity and convenience of the practical application of the formulas we have just found depend on the value of n, and the related successive values of cos kx and sin kx.

TABLE 38. Chart U and Sign Chart M

$$U = \begin{bmatrix} -8,36 & +17,42 & \cdot & \cdot \\ -8,06 & +14,62 & +15,79 & -6,03 \\ -4,29 & +8,87 & +10,76 & -1,29 \\ +2,25 & \cdot & +4,81 & \end{bmatrix} \qquad M = \begin{bmatrix} + & + & + & + \\ + & - & - & + \\ + & - & + & - \\ + & + & - & - \end{bmatrix}$$

TABLE 39. Layout of Computations of the Coefficients of a Periodic-
Type Correlation Equation

I	a_0	a_2	a_4		II	a_1	a_3
+ 9,06	0,5	1	1		−25,78	1,	1
+16,32	0,5	0,5	−0,5		−44,50	0,866	0
+14,05	0,5	−0,5	−0,5		−25,21	0,5	−1
+ 7,06	0,5	−1	1		− 2,56	0	0
6	23,245	3,135	0,935		6	−76,992	−0,570
	3,874	0,523	0,156			−12,832	−0,095

III	b_2	b_4		IV	b_1	b_3
−25,78	0	0		+ 9,06	0	0
− 0,86	0,866	0,866		−3,20	0,5	1
− 1,11	0,866	−0,866		−4,89	0,866	0
+ 7,06	0	0		−2,56	1	−1
6	−1,706	0,216		6	−8,395	−0,640
	−0,284	0,036			−1,399	−0,107

TABLE 40. Values of cos kx and sin kx for x = 0, $\pi/6$, $\pi/3$, ..., $11\pi/6$

x	$\cos x$	$\cos 2x$	$\cos 3x$	$\cos 4x$	$\sin x$	$\sin 2x$	$\sin 3x$	$\sin 4x$
0	1	1	1	1	0	0	0	0
$\frac{\pi}{6}$	0,866	0,5	0	−0,5	0,5	0,866	1	0,866
$\frac{\pi}{3}$	0,5	−0,5	−1	−0,5	0,866	0,866	0	−0,866
$\frac{\pi}{2}$	0	−1	0	1	1	0	−1	0
$\frac{2}{3}\pi$	−0,5	−0,5	1	−0,5	0,866	−0,866	0	0,866
$\frac{5}{6}\pi$	−0,866	0,5	0	−0,5	0,5	−0,866	1	−0,866
π	−1	1	−1	1	0	0	0	0
$\frac{7}{6}\pi$	−0,866	0,5	0	−0,5	−0,5	0,866	−1	0,866
$\frac{4}{3}\pi$	−0,5	−0,5	1	−0,5	−0,866	0,866	0	−0,866
$\frac{3}{2}\pi$	0	−1	0	1	−1	0	1	0
$\frac{5}{3}\pi$	0,5	−0,5	−1	−0,5	−0,866	−0,866	0	0,866
$\frac{11}{6}\pi$	0,866	0,5	0	−0,5	−0,5	−0,866	−1	−0,866

TABLE 41

Months	u	$\tilde{u}$
1	− 8,36	− 8,37
2	− 8,06	− 8,08
3	− 4,29	− 4,28
4	+ 2,25	+ 2,21
5	+ 8,87	+ 8,92
6	+14,62	+14,58
7	+17,42	+17,48
8	+15,79	+15,76
9	+10,76	+10,79
10	+ 4,81	+ 4,79
11	− 1,29	− 1,30
12	− 6,03	− 6,03

Particularly simple is the case when

$$n = 12.$$

In this case, the series of observations may be written in the form

$$\left\{ \begin{array}{cccccccccccc} 0 & \frac{\pi}{6} & \frac{\pi}{3} & \frac{\pi}{2} & \frac{2}{3}\pi & \frac{5}{6}\pi & \pi & \frac{7}{6}\pi & \frac{4}{3}\pi & \frac{3}{2}\pi & \frac{5}{3}\pi & \frac{11}{6}\pi \\ u_0 & u_1 & u_2 & u_3 & u_4 & u_5 & u_6 & u_7 & u_8 & u_9 & u_{10} & u_{11} \end{array} \right\}$$

For laying out the computations, we take into account that, in the four quadrants from 0 to 2π, the cosine and sine assume, four times, the same absolute values, namely, 0, 0.5, 0.866, and 1, sometimes with a plus sign, sometimes with a minus sign. Therefore, before multiplying the data by the successive values of cos kx and sin kx, we can unify these data by quadrant; for example, for b_k

$$u_1 + u_5 - u_7 - u_{11}.$$

We will use an example to demonstrate the sequence of computational operations.

We shall set up a periodic-type correlation equation with terms up to $a_4 \cos 4x$ and $b_4 \sin 4x$ (the influence of the higher harmonics can be ignored) to express the variation of mean monthly air temperature in Leningrad in the century from 1816 to 1915 (Table 37).

First of all, we write the observed data into the columns of chart U, writing these data first downward, then up, then down and then up, marking by dots the places skipped (Table 38).

Then, we add these numbers by row of chart U, adding the same row in four different ways in accordance with the four rows of signs of chart M. We thus obtain four individual groups of totals which are then combined with the cosines and sines of 0, $\pi/6$, $\pi/3$, and $\pi/2$ in the four individual blocks, I, II, III, and IV, shown in Table 39.

For example, the number −25.78, in the first row of block II, is obtained by adding the numbers in the first row of chart U using the signs in the second row of chart M, specifically:

$$+(-8,36) - (+17,42) = -25,78.$$

By then multiplying the numbers in the left-hand column of block II by, respectively,

$$1, \ 0,866, \ 0,5, \ \text{and} \ 0$$

and then adding, we obtain −76.992; by dividing this number by 6, we find

$$a_1 = -12,832.$$

Having thus determined the coefficients of equation (98), we can present the correlation equation expressing the variation in mean monthly air temperature in Leningrad in the form

$$\tilde{u} = 3,874 - 12,832 \cos x + 0,523 \cos 2x - 0,095 \cos 3x + 0,156 \cos 4x -$$

$$- 1,399 \sin x - 0,284 \sin 2x - 0,107 \sin 3x + 0,036 \sin 4x.$$

Giving, in this equation, the variable x the values

$$0, \ \frac{\pi}{6}, \ \frac{\pi}{3}, \ \cdots, \ \frac{11}{6}\pi$$

and using Table 40, we find the probable values, $\tilde{u}$.

From a comparison of the probable values, $\tilde{u}$, with the observed values, u (Table 41), we see that the correlation equation we have obtained very well expresses the variations in mean monthly air temperature in Leningrad. The periodicity of the phenomenon here causes no surprise, since, a priori, each new year must exhibit the same stochastic features as the previous one.

CHAPTER VI

MULTIPLE CORRELATION EQUATIONS

1. Chebyshev's Method for Setting Up Multiple Correlation Equations

To investigate the relationships among several random variables, one sets up multiple correlation equations. As for ordinary correlation equations, Chebyshev's method is a very convenient way of setting up multiple correlation equations.

A linear multiple correlation equation, expressing the functional dependence of one random variable, X_1, on other random variables, has the following form:

$$r_{1|(j_3)|(j_3)|\ldots|(j_n)} = \sum_{g=2}^{n} \frac{R_{g_1}^{(g)} R_{g_1}^{(g)*}}{R_{11}^{(g-1)} R_{11}^{(g)}} \tag{102}$$

with standard error

$$\frac{\sigma_{1.23\ldots n}^2}{\sigma_1^2} = 1 - \sum_{g=2}^{n} \frac{R_{g_1}^{(g)2}}{R_{11}^{(g-1)} R_{11}^{(g)}}. \tag{103}$$

In equation (102),

$$R^{(g)} = \begin{vmatrix} 1 & r_{12} & r_{13} & \ldots & r_{1g} \\ r_{21} & 1 & r_{23} & \ldots & r_{2g} \\ r_{31} & r_{32} & 1 & \ldots & r_{3g} \\ \cdot & \cdot & \cdot & \cdot & \cdot \\ r_{g1} & r_{g2} & r_{g3} & \ldots & 1 \end{vmatrix}, \tag{104}$$

where r_{pq} is the correlation coefficient of the random variables X_p and X_q;

$$R^{(g)*} = \begin{vmatrix} \xi_{1(j_1)} & \xi_{2(j_3)} & \xi_{3(j_3)} & \ldots & \xi_{g(j_g)} \\ r_{21} & 1 & r_{23} & \ldots & r_{2g} \\ r_{31} & r_{32} & 1 & \ldots & r_{3g} \\ \cdot & \cdot & \cdot & \cdot & \cdot \\ r_{g1} & r_{g2} & r_{g3} & \ldots & 1 \end{vmatrix}, \tag{105}$$

and $R_{pq}^{(g)}$ and $R_{pq}^{(g)*}$ are the minors of the determinants of (104) and (105) corresponding to the element r_{pq}. Determinant $R^{(g)}$ is called the correlation determinant.

In increasing successively the number of other random variables, i.e., in increasing n, we can put equation (102) in the form

$$r_{1|(j_3)|(j_3)|(j_4)|\ldots}^{(1,\ldots,1)} = r_{12}\xi_{2(j_3)} + \frac{\begin{vmatrix} r_{12} & r_{13} \\ 1 & r_{23} \\ 1 & r_{23} \\ r_{32} & 1 \end{vmatrix}}{} \cdot \begin{vmatrix} \xi_{2(j_3)} & \xi_{3(j_3)} \\ 1 & r_{23} \end{vmatrix} +$$

$$+ \frac{\begin{vmatrix} r_{12} & r_{13} & r_{14} \\ 1 & r_{23} & r_{24} \\ r_{32} & 1 & r_{34} \\ 1 & r_{23} & r_{34} \\ r_{32} & 1 & r_{34} \\ r_{42} & r_{43} & 1 \end{vmatrix}}{} \cdot \frac{\begin{vmatrix} \xi_{2(j_3)} & \xi_{3(j_3)} & \xi_{4(j_4)} \\ 1 & r_{23} & r_{24} \\ r_{32} & 1 & r_{34} \end{vmatrix}}{\begin{vmatrix} 1 & r_{23} \\ r_{32} & 1 \end{vmatrix}} + \ldots \tag{106}$$

with standard error

$$\frac{\sigma^2_{1.234\ldots}}{\sigma^2_1} = 1 - r^2_{12} - \frac{\begin{vmatrix} r_{12} & r_{13} \\ 1 & r_{23} \end{vmatrix}^2}{\begin{vmatrix} 1 & r_{23} \\ r_{32} & 1 \end{vmatrix}} - \frac{\begin{vmatrix} r_{12} & r_{13} & r_{14} \\ 1 & r_{23} & r_{24} \\ r_{32} & 1 & r_{34} \end{vmatrix}^2}{\begin{vmatrix} 1 & r_{23} \\ r_{32} & 1 \end{vmatrix} \cdot \begin{vmatrix} 1 & r_{23} & r_{24} \\ r_{32} & 1 & r_{34} \\ r_{42} & r_{43} & 1 \end{vmatrix}} - \cdots \tag{107}$$

By retaining only the first term of equation (106), we obtain the ordinary first-order correlation equation, expressing the dependence of X_1 on X_2:

$$r^{(1)}_{1|(j_2)} = r_{12}\xi_{2(j_2)} \tag{108}$$

with standard error

$$\sigma_{1.2} = \sigma_1\sqrt{1 - r^2_{12}} \tag{109}$$

[cf., (20)].

By joining to the first term of equation (106) its second term, we get the multiple linear correlation equation expressing the dependence of X_1 on X_2 and X_3:

$$r^{(1,1)}_{1|(j_2)|(j_3)} = r_{12}\xi_{2(j_2)} + \frac{\begin{vmatrix} r_{12} & r_{13} \\ 1 & r_{23} \end{vmatrix}}{\begin{vmatrix} 1 & r_{23} \\ r_{32} & 1 \end{vmatrix}} \cdot \begin{vmatrix} \xi_{2(j_2)}\xi_{3(j_3)} \\ 1 & r_{23} \end{vmatrix} \tag{110}$$

with standard error

$$\sigma_{1.23} = \sigma_1\sqrt{1 - r^2_{12} - \frac{\begin{vmatrix} r_{12} & r_{13} \\ 1 & r_{23} \end{vmatrix}^2}{\begin{vmatrix} 1 & r_{23} \\ r_{32} & 1 \end{vmatrix}}}. \tag{111}$$

And so forth.

The multiple nonlinear correlation equation expressing the dependence of X_3 on X_1 and X_2 has the following g e n e r a l form:

$$r^{(h_1, h_2)}_{(j_1)|(j_2)|1} = \sum_{g_1=1}^{h_1} \sum_{g_2=1}^{h_2} \frac{Q^{(g_1, g_2)}_{g_1, g_2} Q^{(g_1, g_2)*}_{g_1, g_2}}{Q^{(g_1, g_2)-1}Q^{(g_1, g_2)}}. \tag{112}$$

In this equation,

$$Q^{(g_1, g_2)} = \begin{vmatrix} 1 & 0 & 0 & r_{1|1|0} & 1 & 1 & \ldots & r_{0|g_2|0} \\ 0 & 1 & r_{1|1|0} & r_{2|1|0} & r_{3|0|0} & r_{1|2|0} & \ldots & r_{1|g_2|0} \\ 0 & r_{1|1|0} & 1 & r_{1|2|0} & r_{2|1|0} & r_{0|3|0} & \ldots & r_{0|g_2+1|0} \\ r_{1|1|0} & r_{2|1|0} & r_{1|2|0} & r_{2|2|0} & r_{3|1|0} & r_{1|3|0} & \ldots & r_{1|g_2+1|0} \\ 1 & r_{3|0|0} & r_{2|1|0} & r_{3|1|0} & r_{4|0|0} & r_{2|2|0} & \ldots & r_{2|g_2|0} \\ 1 & r_{1|2|0} & r_{0|3|0} & r_{1|3|0} & r_{2|2|0} & r_{0|4|0} & \ldots & r_{0|g_2+2|0} \\ \cdot & \cdot & \cdot & \cdot & \cdot & \cdot & \ldots & \cdot \\ r_{0|g_2|0} & r_{1|g_2|0} & r_{0|g_2+1|0} & r_{1|g_2+1|0} & r_{2,g_2|0} & r_{0|g_2+2|0} & \ldots & r_{0|2g_2|0} \end{vmatrix} \tag{113}$$

where $r_{g_1|g_2|0}$ is the empirical mixed standard moment of the variables X_1 and X_2; $Q^{(g_1, g_2)-1}$ is the determinant one order lower than the determinant of (113); $Q^{(g_1, g_2)}_{g_1, g_2}$ are the determinants formed from the determinant of (113) by successively replacing the elements of its last column by the moments

$$0, \ r_{1|0|1}, \ r_{0|1|1}, \ r_{1|1|1}, \ r_{2|0|1}, \ r_{0|2|1}, \ \ldots, \ r_{0|g_2|1},$$

and, finally, $Q^{(g_1, g_2)^*}_{g_1, g_2}$ are the determinants formed from the determinant in (113) by successively replacing the elements of its last column by the quantities

$$1, \; \xi_{1(j_1)}, \; \xi_{2(j_2)}, \; \xi_{1(j_1)} \xi_{2(j_2)}, \; \xi^2_{1(j_1)}, \; \xi^2_{2(j_2)}, \; \ldots , \; \xi^{g_2}_{2(j_2)}.$$

The standard error of the multiple nonlinear correlation equation of (112) equals

$$\frac{\sigma^{(h_1, h_2)2}_{3.12}}{\sigma^2_3} = 1 - \sum_{g_1=1}^{h_1} \sum_{g_2=1}^{h_2} \frac{Q^{(g_1, g_2)2}_{g_1, g_2}}{Q^{(g_1, g_2)-1} Q^{(g_1, g_2)}}. \tag{114}$$

To simplify the calculation of correlation equation (112), we put it in a somewhat different form. To this end, we introduce the notation

$$\left. \begin{array}{ll} \gamma_1 = 1 - r^2_{1|1|0}, & \gamma_6 = r_{4|0|0} - r^2_{3|0|0} - 1, \\ \gamma_2 = r_{1|2|0} - r_{2|1|0} r_{1|1|0}, & \gamma_7 = r_{0|3|0} - r_{1|2|0} r_{1|1|0}, \\ \gamma_3 = r_{2|2|0} - r^2_{2|1|0} - r^2_{1|1|0}, & \gamma_8 = r_{1|3|0} - r_{2|1|0} r_{1|2|0} - r_{1|1|0}, \\ \gamma_4 = r_{2|1|0} - r_{1|1|0} r_{3|0|0}, & \gamma_9 = r_{2|2|0} - r_{1|2|0} r_{3|0|0} - 1, \\ \gamma_5 = r_{3|1|0} - r_{2|1|0} r_{3|0|0} - r_{1|1|0}, & \gamma_{10} = r_{0|4|0} - r^2_{1|2|0} - 1; \end{array} \right\} \tag{115}$$

$$\left. \begin{array}{l} \delta_1 = r_{0|1|1} - r_{1|1|0} r_{1|0|1}, \\ \delta_2 = r_{1|1|1} - r_{2|1|0} r_{1|0|1}, \\ \delta_3 = r_{2|0|1} - r_{3|0|0} r_{1|0|1}, \\ \delta_4 = r_{0|2|1} - r_{1|2|0} r_{1|0|1}; \end{array} \right\} \tag{116}$$

$$\left. \begin{array}{ll} c_1 = \gamma_1 \gamma_3 - \gamma^2_2, & c_4 = \gamma_1 \gamma_8 - \gamma_2 \gamma_7, \\ c_2 = \gamma_1 \gamma_5 - \gamma_2 \gamma_4, & c_5 = \gamma_1 \gamma_9 - \gamma_4 \gamma_7, \\ c_3 = \gamma_1 \gamma_6 - \gamma^2_4, & c_6 = \gamma_1 \gamma_{10} - \gamma^2_7; \end{array} \right\} \tag{117}$$

$$\left. \begin{array}{l} d_1 = \gamma_1 \delta_2 - \gamma_2 \delta_1, \\ d_2 = \gamma_1 \delta_3 - \gamma_4 \delta_1, \\ d_3 = \gamma_1 \delta_4 - \gamma_7 \delta_1. \end{array} \right\} \tag{118}$$

By employing this notation, and limiting ourselves to the first five terms of expression (112) for the cases when $h_1 \le 2$, $h_2 \le 2$, we can bring the multiple correlation equation to the form

$$r^{(2,2)}_{(j_1)|(j_2)|1} = r_{1|0|1} \xi_{1(j_1)} + \frac{\delta_1}{\gamma_1} (\xi_{2(j_2)} - r_{1|1|0} \xi_{1(j_1)}) +$$

$$+ \frac{d_1}{c_1} \left[\xi_{1(j_1)} \xi_{2(j_2)} - r_{2|1|0} \xi_{1(j_1)} - r_{1|1|0} - \frac{\gamma_2}{\gamma_1} (\xi_{2(j_2)} - r_{1|1|0} \xi_{1(j_1)}) \right] +$$

$$+ \frac{\begin{vmatrix} c_1 & d_1 \\ c_2 & d_2 \\ c_1 & c_2 \\ c_2 & c_3 \end{vmatrix}}{} \left\{ \xi^2_{1(j_1)} - r_{3|0|0} \xi_{1(j_1)} - 1 - \frac{\gamma_4}{\gamma_1} (\xi_{2(j_2)} - r_{1|1|0} \xi_{1(j_1)}) - \right.$$

$$\left. - \frac{c_2}{c_1} \left[\xi_{1(j_1)} \xi_{2(j_2)} - r_{2|1|0} \xi_{1(j_1)} - r_{1|1|0} - \frac{\gamma_2}{\gamma_1} (\xi_{2(j_2)} - r_{1|1|0} \xi_{1(j_1)}) \right] \right\} +$$

55

$$+ \frac{\begin{vmatrix} c_1 & c_2 & d_1 \\ c_2 & c_3 & d_2 \\ c_4 & c_5 & d_3 \end{vmatrix}}{\begin{vmatrix} c_1 & c_2 & c_4 \\ c_2 & c_3 & c_5 \\ c_4 & c_5 & c_6 \end{vmatrix}} \left(\xi_2^2 {}_{(j_2)} - r_{1|2|0} \xi_1 {}_{(j_1)} - 1 - \frac{\gamma_2}{\gamma_1} (\xi_2 {}_{(j_2)} - r_{1|1|0} \xi_1 {}_{(j_1)}) - \right.$$

$$- \frac{c_4}{c_1} \left[\xi_1 {}_{(j_1)} \xi_2 {}_{(j_2)} - r_{2|1|0} \xi_1 {}_{(j_1)} - r_{1|1|0} - \frac{\gamma_2}{\gamma_1} (\xi_2 {}_{(j_2)} - r_{1|1|0} \xi_1 {}_{(j_1)}) \right] -$$

$$- \frac{\begin{vmatrix} c_1 & c_4 \\ c_2 & c_5 \end{vmatrix}}{\begin{vmatrix} c_1 & c_2 \\ c_2 & c_3 \end{vmatrix}} \left\{ \xi_1^2 {}_{(j_1)} - r_{3|0|0} \xi_1 {}_{(j_1)} - 1 - \frac{\gamma_4}{\gamma_1} (\xi_2 {}_{(j_2)} - r_{1|1|0} \xi_1 {}_{(j_1)}) - \right.$$

$$\left. \left. - \frac{c_2}{c_1} \left[\xi_1 {}_{(j_1)} \xi_2 {}_{(j_2)} - r_{2|1|0} \xi_1 {}_{(j_1)} - r_{1|1|0} - \frac{\gamma_2}{\gamma_1} (\xi_2 {}_{(j_2)} - r_{1|1|0} \xi_1 {}_{(j_1)}) \right] \right\} \right). \tag{119}$$

The standard error of this equation is

$$\frac{\sigma_{3 \cdot 12}^{(2,2)\,2}}{\sigma_3^2} = 1 - r_{1|0|1}^2 - \frac{\delta_1^2}{\gamma_1 c_1} - \frac{\begin{vmatrix} c_1 & d_1 \\ c_3 & d_2 \end{vmatrix}^2}{c_1 \cdot \begin{vmatrix} c_1 & c_2 \\ c_2 & c_3 \end{vmatrix}} - \frac{\begin{vmatrix} c_1 & c_2 & d_1 \\ c_2 & c_3 & d_2 \\ c_4 & c_5 & d_3 \end{vmatrix}^2}{\begin{vmatrix} c_1 & c_2 \\ c_2 & c_3 \end{vmatrix} \cdot \begin{vmatrix} c_1 & c_2 & c_4 \\ c_2 & c_3 & c_5 \\ c_4 & c_5 & c_6 \end{vmatrix}}. \tag{120}$$

If we restrict ourselves to the first term of equation (119), we obtain the ordinary linear correlation equation expressing the dependence of X_3 on X_1 [cf., (108)]. By adding the second term, we get the multiple linear correlation equation expressing the dependence of X_3 on X_1 and X_2 [cf., (110)]. If, after this, we add the third term, containing the product $\xi_{1(j_1)} \xi_{2(j_2)}$, we obtain a multiple correlation equation of the hyperbolic type. And so forth.

The transition from the "approximate" conditional standard moments, $r_{(j_1)|(j_2)|1}^{(h_1, h_2)}$ to the probable values, $\tilde{X}_{(j_1)|(j_2)|1}$, is made via the formula

$$\tilde{X}_{(j_1)|(j_2)|1} = \bar{X}_3 + r_{(j_1)|(j_2)|1}^{(h_1, h_2)} \sigma_3 \tag{121}$$

[cf., (14)].

2. Computation of Multiple Correlation Equations

As an example of the calculation of multiple correlation equations, we shall consider the correlation relationship between weight-by-volume, X_1 (γ, mg/cm^3), impact toughness, X_2 (H_{rad}, gmm/mm^2), and yield strength with compression along the grain, X_3 (σ_B, kg/cm^2), of birch wood. The test pieces were cubes measuring $3 \times 3 \times 3$ cm. Three such cubes each were taken from each sample to be tested, and the averages of each such triplet were used as the observed values of the quantities under study.

We shall set up the correlation equation expressing the dependence of yield strength, upon compression with the grain of birch, on weight-by-volume and impact toughness.

Limiting ourselves, for this, to the first three terms of equation (119), we must compute the empirical means and standard deviations of the variables in question, as well as their standard moments

$$r_{1|1|0}, \quad r_{1|0|1}, \quad r_{0|1|1}, \quad r_{2|1|0}, \quad r_{1|2|0}, \quad r_{2|2|0}, \quad r_{1|1|1}.$$

The test results, and all the necessary computations are given in Table 42.

TABLE 42. Layout of the Computation of Multiple Correlations for Small Samples

№	X_1	X_2	X_3	x_1	x_2	x_3	x_1^2	x_2^2	x_3^2	x_1x_2	x_1x_3	x_2x_3	$x_1^2x_3$	$x_1x_2^2$	$x_1^2x_2^2$	$x_1x_2x_3$
(1)	(2)	(3)	(4)	(5)	(6)	(7)	(8)	(9)	(10)	(11)	(12)	(13)	(14)	(15)	(16)	(17)
1	648	901	852	+68	+126	+113	4 624	15 876	12 769	8 568	7 684	14 238	582 624	1 079 568	73 410 624	968 184
2	628	828	800	+48	+53	+61	2 304	2 809	3 721	2 544	2 928	3 233	122 112	134 832	6 471 936	155 184
3	511	681	616	−69	−94	−123	4 761	8 836	15 129	6 486	8 487	11 562	−447 534	−609 684	42 068 196	−797 778
4	535	724	693	−45	−51	−41	2 025	2 601	1 681	2 295	1 845	2 091	−103 275	−117 045	5 267 025	−94 095
5	574	822	733	−6	+47	−6	36	2 209	36	−282	36	−282	1 692	−13 254	79 524	1 692
6	580	782	701	0	+7	−38	0	49	1 444	0	0	−266	0	0	0	0
7	654	904	865	+74	+129	+126	5 476	16 641	15 876	9 546	9 324	16 254	706 404	1 231 434	91 126 116	1 202 796
8	576	738	727	−4	−37	−12	16	1 369	144	148	48	444	−592	−5 476	21 904	−1 776
9	508	712	562	−72	−63	−177	5 184	3 969	31 329	4 536	12 744	11 151	−326 592	−285 768	20 575 296	−802 872
10	572	760	726	−8	−15	−13	64	225	169	120	104	195	−960	−1 800	14 400	−1 560
11	536	695	681	−44	−80	−58	1 936	6 400	3 364	3 520	2 552	4 640	−154 880	−281 600	12 390 400	−204 160
12	618	793	819	+38	+18	+80	1 444	324	6 400	684	3 040	1 440	25 992	+12 312	467 856	54 720
13	571	746	721	−9	−29	−18	81	841	324	261	162	522	−2 349	−7 569	68 121	−4 698
14	540	749	675	−40	−26	−64	1 600	676	4 096	1 040	2 560	1 664	−41 600	−27 040	1 081 600	−66 560
15	615	774	833	+35	−1	+94	1 225	1	8 836	−35	3 290	−94	−1 225	35	1 225	−3 290
16	622	782	817	+42	+7	+78	1 764	49	6 084	294	3 276	546	12 348	2 058	86 436	22 932
—	—	—	—	−297 / +305	−396 / +387	−550 / +552	—	—	—	−317 / +40 042	− / +58 080	−642 / +67 980	−1 079 007 / +1 451 172	−1 349 236 / +2 460 239	− / −	−1 976 789 / +2 405 508
Σ	9 288	12 391	11 821	+8	−9	+2	32 540	62 875	111 402	+39 725	+58 080	+67 338	+372 165	+1 111 003	253 130 659	+428 719
—	$\bar X_1 =$ $= 580$	$\bar X_2 =$ $= 775$	$\bar X_3 =$ $= 739$	$m_{1000} =$ $= +0,5$	$m_{0100} =$ $= -0,56$	$m_{0010} =$ $= +0,125$	$\mu_{2000} =$ $= 2034$	$\mu_{0200} =$ $= 3930$	$\mu_{0020} =$ $= 6963$	$\mu_{1100} =$ $= 2483$	$\mu_{1010} =$ $= 3630$	$\mu_{0110} =$ $= 4209$	$\mu_{2010} =$ $= 23260$	$\mu_{1200} =$ $= 69438$	$\mu_{2200} =$ $= 15820666$	$\mu_{1110} =$ $= 26795$
—							$\sigma_1 =$ $= 45$	$\sigma_2 = 63$	$\sigma_3 = 83$	$r_{1100} =$ $= 0,876$	$r_{1010} =$ $= 0,971$	$r_{0110} =$ $= 0,805$	$r_{2010} =$ $= 0,182$	$r_{1200} =$ $= 0,393$	$r_{2200} =$ $= 1,979$	$r_{1110} =$ $= 0,114$

57

TABLE 43

№	σ_B	σ_B^I	σ_B^{II}	σ_B^{III}	(2) — (3)	(2) — (4)	(2) — (5)
(1)	(2)	(3)	(4)	(5)	(6)	(7)	(8)
1	852	861	850	848	− 9	+ 2	+ 4
2	800	825	826	827	− 25	− 26	− 27
3	616	615	618	615	+ 1	− 2	+ 1
4	698	658	657	657	+ 40	+ 41	+ 41
5	733	728	714	717	+ 5	+ 19	+ 16
6	701	739	737	739	− 38	− 36	− 38
7	865	871	861	859	− 6	+ 4	+ 6
8	727	732	740	741	− 5	− 13	− 14
9	562	610	603	603	− 48	− 41	− 41
10	726	724	715	727	+ 2	+ 11	− 1
11	681	660	666	666	+ 21	+ 15	+ 15
12	819	807	814	815	+ 12	+ 5	+ 4
13	721	723	727	728	− 2	− 6	− 7
14	675	667	661	663	+ 8	+ 14	+ 12
15	833	801	813	814	+ 32	+ 20	+ 19
16	817	814	826	826	+ 3	− 9	− 9
Σ	11 826	11 835	11 828	11 845	− 9	− 2	− 19

Column (1) of this table gives the ordinal numbers of the tests, while columns (2)–(4) provide the values of the variables in question. Adding the numbers in each of these three columns, and then dividing the sums by the size of the sample, n = 16, we find the means

$$\bar{X}_1 = 580 \text{ mg/cm}^3, \quad \bar{X}_2 = 775 \text{ gmm/mm}^2, \quad \bar{X}_3 = 739 \text{ kg/cm}^2.$$

In columns (5)–(7) we enter the deviations, x_1, x_2, and x_3, of the observed values of the random variables from their means. Since the computed means were rounded to the nearest integer, the sums of columns (5)–(7), divided by the sample size (giving the first initial moments), will show the error admitted in holding to the given precision of the computations. We have

$$m_{1|0|0} = +0.5, \quad m_{0|1|0} = -0.56, \quad m_{0|0|1} = +0.125;$$

and the relative errors will equal, respectively,

$$\varepsilon_1 = \frac{+0.5}{580} = +0.0009, \quad \varepsilon_2 = \frac{0.56}{775} = -0.0008,$$
$$\varepsilon_3 = \frac{+0.125}{739} = +0.0002,$$

i.e., will in all cases be less than 0.001; such errors may be neglected.

The make-up of the remaining columns of Table 42 is clear from the column headings. Since x_1, x_2, and x_3 are deviations from means, the sums of each column, divided by sample size, give the corresponding central moments.

Taking the square root of the second central moments [columns (8)–(10)], we find the standard deviations

$$\sigma_1 = 45 \text{ mg/cm}^3, \quad \sigma_2 = 63 \text{ gmm/mm}^2, \quad \sigma_3 = 83 \text{ kg/cm}^2.$$

We obtain the mixed standard moments from the mixed central moments by the formula

$$r_{h_1 \mid h_2 \mid h_3} = \frac{\mu_{h_1 \mid h_2 \mid h_3}}{\sigma_1^{h_1} \sigma_2^{h_2} \sigma_3^{h_3}}.$$

We then have [columns (11)–(17)]:

$$r_{1 \mid 1 \mid 0} = 0.876, \quad r_{1 \mid 0 \mid 1} = 0.971, \quad r_{0 \mid 1 \mid 1} = 0.805,$$
$$r_{2 \mid 1 \mid 0} = 0.182, \quad r_{1 \mid 2 \mid 0} = 0.393, \quad r_{2 \mid 2 \mid 0} = 1.979,$$
$$r_{1 \mid 1 \mid 1} = 0.114.$$

Using these moments, we find the constants γ, δ, c, and d, from formulas (115)–(118):

$$\gamma_1 = +0.233, \quad \gamma_2 = +0.234, \quad \gamma_3 = +1.178,$$
$$\delta_1 = -0.046, \quad \delta_2 = -0.063,$$
$$c_1 = +0.220, \quad d_1 = -0.004.$$

By then substituting these constants in (119), we obtain the correlation equation in some form or another.

In particular, the ordinary linear correlation equation expressing the dependence of yield strength of birch for compression with the grain, X_3, on weight-by-volume, X_1, has the form

$$r^{(1)}_{(j_1) \cdot \mid 1} = 0.971 \, \xi_{1 \, (j_1)}.$$

Going to the units of measurement [cf., (23)], we find

$$\tilde{X}_{(j_1) \mid \cdot \mid 1} = 739 + 0.971 \cdot \frac{X_{1 \, (j_1)} - 580}{45} \cdot 83 = -300 + 1.791 X_{1 \, (j_1)} \tag{122}$$

with standard error

$$\sigma^{(1)}_{3 \cdot 1} = 198.507 \ \text{kg/cm}^2.$$

The multiple linear correlation equation expressing the dependence of yield strength, X_3, on weight-by-volume, X_1, and on impact toughness, X_2, has the form

$$r^{(1; 1)}_{(j_1) \mid (j_2) \mid 1} = 0.971 \xi_{1 \, (j_1)} - 0.197 \, (\xi_{2 \, (j_2)} - 0.876 \xi_{1 \, (j_1)}) = 1.444 \xi_{1 \, (j_1)} - 0.197 \xi_{2 (j_2)}$$

or

$$\tilde{X}_{(j_1) \mid (j_2) \mid 1} = 739 + \left\{ 0.971 \cdot \frac{X_{1 \, (j_1)} - 580}{45} \quad 0.197 \left(\frac{X_{2 \, (j_2)} - 775}{63} - 0.876 \cdot \frac{X_{1 \, (j_1)} - 580}{45} \right) \right\} \cdot 83 =$$
$$= -283 + 2.109 \, X_{1 \, (j_1)} - 0.260 \, X_{2 \, (j_2)} \tag{123}$$

with standard error

$$\sigma^{(1; 1)}_{3 \cdot 12} = 182.033 \ \text{kg/cm}^2.$$

Finally, the multiple correlation equation of the hyperbolic type turned out to be the following:

$$r_{(j_1) \mid (j_2) \mid 1} = 0.971 \xi_{1 \, (j_1)} - 0.197 \, (\xi_{2 \, (j_2)} - 0.876 \xi_{1 \, (j_1)}) -$$
$$- 0.0182 \, [\xi_{1 \, (j_1)} \xi_{2 \, (j_2)} - 0.182 \xi_{1 \, (j_1)} - 0.876 -$$
$$- 1.0043 \, (\xi_{2 \, (j_2)} - 0.876 \xi_{1 \, (j_1)})]$$

or

$$\tilde{X}_{(j_1) \mid (j_2) \mid 1} = -511.4693 + 2.4728 X_{1 \, (j_1)} + 0.0546 X_{2 \, (j_2)} - 0.0005 X_{1 \, (j_1)} X_{2 \, (j_2)} \tag{124}$$

with standard error

$$\sigma'_{3.12} = 181.465 \ \text{kg/cm}^2.$$

The probable values of yield strength of birch upon compression with the grain, σ_B^I, σ_B^{II}, and σ_B^{III}, computed from these equations, as well as the deviations of the computed from the observed values, are given in Table 43.

CHAPTER VII

DISTRIBUTION SURFACES

1. Normal Distribution Surface

The final goal in investigating relationships between random variables is the establishment of the equation of the corresponding distribution surface. Here, we shall consider the normal distribution surface, and its generalization — the type A distribution surface.

The equation of the bivariate normal distribution surface has the form

$$f(x_1, \ x_2) = \frac{1}{2\pi\sigma_1\sigma_2\sqrt{1-r^2}} e^{-\frac{1}{2(1-r^2)}\left[\left(\frac{x_1}{\sigma_1}\right)^2 - 2r\frac{x_1}{\sigma_1}\cdot\frac{x_2}{\sigma_2} + \left(\frac{x_2}{\sigma_2}\right)^2\right]}, \tag{125}$$

where x_1 and x_2 are the deviations from the corresponding means.

For $x_1 = x_2 = 0$, i.e., at the point corresponding to the center of the distribution, the function $f(x_1, x_2)$ has its maximum value; as the absolute values of x_1 and x_2 increase, the value of the function $f(x_1, x_2)$ decreases, initially very rapidly, then more slowly, tending to zero with unbounded increase of the absolute values of x_1 and x_2 (Fig. 8).

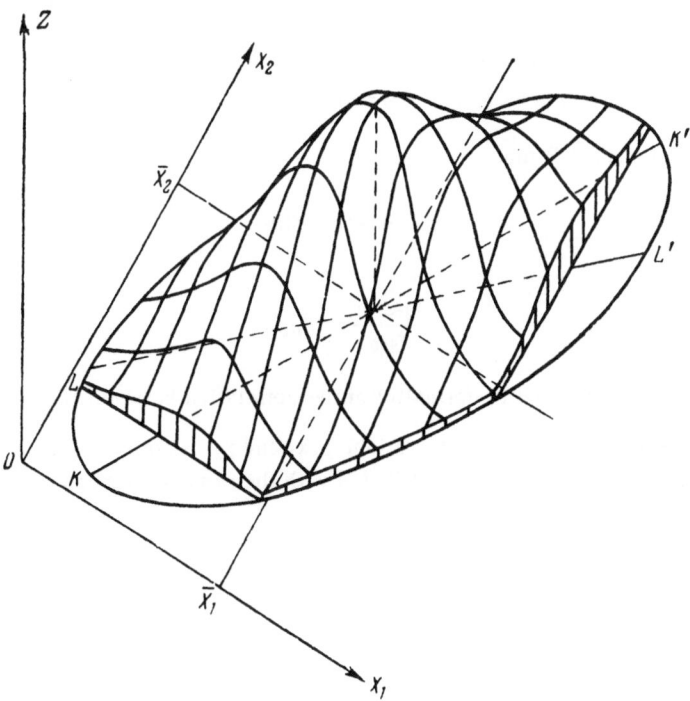

Fig. 8. The normal distribution surface, its vertical sections, and the correlation lines.

TABLE 44. Statistics for Table 1

h_1 \ h_2	$r_{h_1\mid h_2}$ 0	1	2	3	4
0	1	0	1	$-0,036$	3,113
1	0	$+0,880$	$+0,005$	2,658	
2	1	$+0,051$			
3	$+0,101$	2,541			
4	2,893				

$X_{1(a)} = 0,445$,	$X_{2(a)} = 130\,000$ kg/cm²,
$m_{1\cdot0} = +0,335$,	$m_{0\cdot1} = -0,082$,
$\bar{X}_1 = 0,455 \pm 0,091$ g/cm³,	$\bar{X}_2 = 128852 \pm 1134$ kg/cm²
$\sigma_1 = 1,824 \pm 0,064$.	$\sigma_2 = 1,620 \pm 0,058$.

$r = +0,880 \pm 0,018$, $\eta_{21} = \eta_{12} = 0,882 \pm 0,016$,

$\sqrt{1-r^2} = 0,47497$.

The following relationships subsit among the moments of the normal distribution surface:

$$r_{2s\mid2t+1} = r_{2s+1\mid2t} = 0, \tag{126}$$

$$r_{2t\mid0} = r_{0\mid2t} = (2t-1)!!, \tag{127}$$

$$r_{2t+1\mid1} = r_{1\mid2t+1} = (2t+1)!!\, r_{1\mid1}. \tag{128}$$

In particular

$$\left.\begin{array}{l} r_{0\mid3} = r_{3\mid0} = 0, \quad r_{1\mid2} = r_{2\mid1} = 0, \quad r_{0\mid1} = r_{4\mid0} = 3, \\ r_{1\mid3} = r_{3\mid1} = 3r_{1\mid1}, \quad r_{2\mid2} = 1 + 2r_{1\mid1}^2. \end{array}\right\} \tag{129}$$

It follows from (127) and (128) that

$$r_{2t-1\mid1} = r_{2t\mid0}\, r_{1\mid1}. \tag{130}$$

From whence,

$$\frac{r_{2t-1\mid1}}{r_{2t\mid0}} = r_{1\mid1}. \tag{131}$$

The equality may serve as the criterion of normality of the correlation between two random variables.

If the joint distribution of the two random variables, x_1 and x_2, is normal, with zero means, then each of these variables, taken separately, is normally distributed with the same zero mean, and with standard deviations of σ_1 and σ_2, respectively.

We now set, in (125),

$$x_1 = X_1,$$

where X_1 is arbitrary. We then obtain

$$f(X_1, x_2) = \frac{1}{2\pi\sigma_1\sigma_2\sqrt{1-r^2}}\, e^{-\frac{1}{2(1-r^2)}\left[\left(\frac{X_1}{\sigma_1}\right)^2 - 2r\frac{X_1}{\sigma_1}\cdot\frac{x_2}{\sigma_2} + \left(\frac{x_2}{\sigma_2}\right)^2\right]}. \tag{132}$$

TABLE 45. Normalized and Observed Frequencies of Table 1

ξ_1 \ ξ_2	-3,652	-3,035	-2,418	-1,800	-1,184	-0,566	+0,051	+0,668	+1,285	+1,902	+2,519	+3,137	+3,754	Σ
-3,473		0,1 —	0,1 —											0,2 —
-2,925		0,4 —	0,6 —	0,2 —										1,2 —
-2,377	0,1 —	0,4 (1)	2,1 (3)	2,2 (2)	0,4 —									5,2 (6)
-1,828		0,1 (1)	2,0 (1)	7,9 (9)	5,7 (6)	0,8 (1)								16,5 (18)
-1,280			0,5 (1)	7,3 (8)	19,9 (20)	10,0 (8)	0,9 —							38,6 (37)
-0,732				1,8 (1)	18,2 (11)	34,2 (42)	11,9 (14)	0,8 —						66,9 (68)
-0,184				0,1 —	4,4 (7)	31,0 (26)	40,4 (46)	9,7 (11)	0,4 —					86,0 (90)
+0,365					0,3 —	7,4 (9)	36,1 (31)	32,5 (36)	5,4 (3)	0,2 —				81,9 (79)
+0,913						0,5 (1)	8,6 (9)	28,7 (23)	17,9 (19)	2,1 (2)				57,8 (54)
+1,461							0,5 —	6,7 (8)	15,6 (15)	6,7 (6)	0,6 (2)			30,1 (31)
+2,009								0,4 (1)	3,6 (4)	5,8 (6)	1,7 (1)	0,1 —		11,6 (12)
+2,558									0,2 (1)	1,3 (2)	1,5 (1)	0,3 —		3,3 (4)
+3,106											0,3 —	0,3 (1)	0,1 —	0,7 (1)
Σ	0,1 —	1,0 (2)	5,3 (5)	19,5 (20)	48,9 (44)	83,9 (87)	98,4 (100)	78,8 (79)	43,1 (42)	16,1 (16)	4,1 (4)	0,7 (1)	0,1 —	400,0 (400)

Completing the square in the exponent, we find

$$f(X_1,\ x_2) = \frac{1}{2\pi\sigma_1\sigma_2\sqrt{1-r^2}}\, e^{-\frac{X_1^2}{2\sigma_1^2} - \frac{1}{2(1-r^2)}\left[\frac{x_2}{\sigma_2} - r\frac{X_1}{\sigma_1}\right]^2} \qquad (133)$$

We now divide (133) by

$$f(X_1) = \frac{1}{\sigma_1\sqrt{2\pi}}\, e^{-\frac{X_1^2}{2\sigma_1^2}},$$

i.e., by the probability density of the first random variable. We obtain

$$f_{X_1}(x_2) = \frac{1}{\sigma_{2.1}\sqrt{2\pi}}\, e^{-\frac{(x_2 - \tilde{x}_2)^2}{2\sigma_{2.1}^2}},$$

(134)

where

$$\tilde{x}_2 = r\frac{\sigma_2}{\sigma_1}X_1,$$

(135)

$$\sigma_{2.1} = \sigma_2\sqrt{1 - r^2}.$$

(136)

Similarly, substituting in (125) $x_2 = X_2$, we find

$$f_{X_2}(x_1) = \frac{1}{\sigma_{1.2}\sqrt{2\pi}}\, e^{-\frac{(x_1 - \tilde{x}_1)^2}{2\sigma_{1.2}^2}},$$

(137)

where

$$\tilde{x}_1 = r\frac{\sigma_1}{\sigma_2}X_2,$$

(138)

$$\sigma_{1.2} = \sigma_1\sqrt{1 - r^2}.$$

(139)

Equation (134) represents the conditional distribution of the random variable x_2, given that random variable x_1 has the value X_1. As is obvious from equation (134), this conditional distribution is normal, with mean of $\tilde{x}_2$ and standard deviation $\sigma_{2.1}$.

Similarly, equation (137) is the conditional distribution of random variable x_1, given that random variable x_2 has value X_2. This distribution is normal with mean $\tilde{x}_1$ and standard deviation $\sigma_{1.2}$.

Equations (135) and (138) are stochastic correlation equations which express the relationship between the variables x_1 and x_2. It is obvious from these equations that, in the case of normal correlation between the variables, the correlation equations are linear. The lines of the conditional means, KK' and LL', on Fig. 8 are the graphs of the correlation equations, and are straight lines intersecting at the point corresponding to the center of the distribution.

We now present the formulas to be used in computing the normalized frequencies for a normal distribution table.* For this, we use tables of values of the normal probability distribution density function.†

We give the normalized observations the values

$$\xi_1 = \frac{x_1}{\sigma_1}, \quad \xi_2 = \frac{x_2}{\sigma_2},$$

(140)

where x_1, x_2, σ_1, and σ_2 are in units of groups (classes), and we introduce the notation

$$x = \xi_1, \quad y = \frac{\xi_2 - r\xi_1}{\sqrt{1 - r^2}},$$

(141)

giving us

$$\tilde{n}_{j_1|j_2} = \frac{n}{\sigma_1\sigma_2\sqrt{1 - r^2}}\, f(x)f(y),$$

(142)

where $f(x)$ and $f(y)$ are normal probability distribution density functions.

---

*That is, tables set up on the basis of observed results of a two-dimensional normal population.

†Such tables are found, e.g., in Hoel, "Introduction to Mathematical Statistics" (Wiley & Sons, 1947, New York), or Kenney, "Mathematics of Statistics" (Van Nostrand, Princeton, New Jersey).

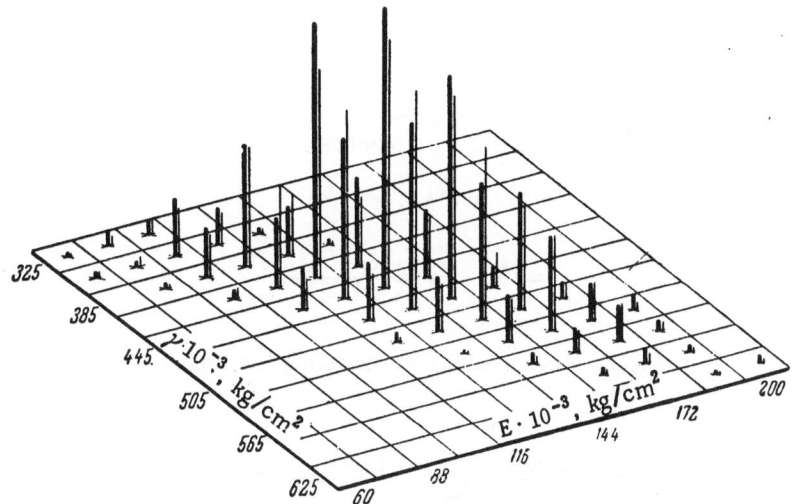

Fig. 9. Observed and computed frequencies of the distribution table of weight-by-volume at 10% moisture content, and modulus of elasticity with compression with the grain of spruce. (Thick lines are observed, thin lines are computed, frequencies.)

As an example, we consider the distribution table of weight-by-volume, at 10% moisture content, and modulus of elasticity, for compression with the grain, of spruce (Table 1) (in Chapter I). The statistics for this table are given here in Table 44. The normalized observed values, ξ_1 and ξ_2, of the quantities γ and E, are given in Table 45.

On the basis of these moments, we find

$$\frac{r_{3|1}}{r_{4|0}} = \frac{2,541}{2,893} = 0,878, \qquad \frac{r_{1|3}}{r_{0|4}} = \frac{2,658}{3,113} = 0,854,$$

i.e., condition (131) is satisfied quite well.

Turning now to the computation of the normalized frequencies for our distribution table, we first determine the coefficients in formula (142):

$$\frac{n}{\sigma_1 \sigma_2 \sqrt{1-r^2}} = \frac{400}{1,824 \cdot 1,620 \cdot 0,47497} = 285,0.$$

To calculate each frequency, $\widetilde{n}_{j_1|j_2}$, we must perform the following acts.

We compute, for example, the frequency $\widetilde{n}_{j_1|j_2}$ for the cell of the table with labels

$$\xi_1 = -1.280, \qquad \xi_2 = -1.184.$$

We have

$$x = -1,280,$$
$$y = \frac{-1,184 - 0,880 \cdot (-1,280)}{0,47497} = -0,122.$$

From a table of $f(x)$ we find

$$f(-1,280) = 0.17585, \quad f(-0,122) = 0,39598.$$

Consequently,

$$f(x) \cdot f(y) = 0.06963.$$

Finally,

$$\tilde{n}_{j_1|j_2} = 285.0 \cdot 0.06963 = 19.9.$$

To check our calculations, we set, instead of (141),

$$y = \xi_2, \qquad x = \frac{\xi_1 - r\xi_2}{\sqrt{1-r^2}}. \tag{143}$$

Then, for the cell of our distribution table we have been considering, we find

$$y = -1.184,$$
$$x = \frac{-1.280 - 0.880 \cdot (-1.184)}{0.47497} = -0.501,$$
$$f(-1.184) = 0.19792, \quad f(-0.501) = 0.35189,$$
$$f(y) \cdot f(x) = 0.06965,$$
$$\tilde{n}_{j_1|j_2} = 285.0 \cdot 0.06965 = 19.9.$$

Computation of $\tilde{n}_{j_1}|_{j_2}$ may also be performed using tables of values of the function $f^*(x)$.

[Trans. note to the reader: These are ordinates of the normal probability curve, erected at distances x/σ from the mean, expressed as decimal fractions of the maximum ordinate. Such tables may be found, e.g., in Croxton and Cowden, "Applied General Statistics" (Prentice-Hall, New York). ESS]

In this case,

$$\tilde{n}_{j_1|j_2} = \tilde{n}_{0|0} f^*(x) f^*(y), \tag{144}$$

where

$$\tilde{n}_{0|0} = \frac{n}{2\pi\sigma_1\sigma_2\sqrt{1-r^2}}. \tag{145}$$

In our example,

$$\tilde{n}_{0|0} = \frac{400}{6.2832 \cdot 1.824 \cdot 1.620 \cdot 0.47497} = 45.4.$$

For the distribution table cell with labels

$$\xi_1 = -1.280, \qquad \xi_2 = -1.184$$

we have

$$f^*(-1.280) = 0.44078, \quad f^*(-0.122) = 0.99259,$$
$$f^*(x) f^*(y) = 0.43752,$$
$$\tilde{n}_{j_1|j_2} = 45.4 \cdot 0.43752 = 19.9.$$

Proceeding in the manner described, we obtain Table 45 (where the observed values are placed in parentheses).

The numbers in the total row and column of the distribution table constitute the normalized frequencies of the distribution groups of each random variable separately, i.e., the marginal distributions. A comparison of these numbers with the frequencies computed from the equations

$$\tilde{n}_{j_1\cdot} = 87.5 f^*(\xi_1) \quad \text{and} \quad \tilde{n}_{\cdot|j_2} = 98.48 f^*(\xi_2)$$

indicates that our computations were correctly performed.

TABLE 46

D, cm \ H, m	17,5	18,5	19,5	20,5	21,5	22,5	23,5	24,5	25,5	26,5	27,5	28,5	29,5	Σ
16	1	1	3											5
20		1	2	4	6	4	2	1						20
24			1	2	7	10	12	7	4					43
28				3	10	14	19	14	7	1	1			69
32					3	12	21	22	20	8	2			88
36						1	3	15	24	20	12	4		79
40							2	2	11	21	16	3		55
44								1	3	12	8	6	1	31
48									1	1	3	4	2	1	12
52											1	2	1	4
56												1	1	2
Σ	1	2	6	6	16	28	45	67	79	84	52	20	2	408

The frequencies shown in the cells of Table 45 are presented graphically on Fig. 9. Comparing these frequencies, we see a good match between them.

2. Type A Distribution Surface

The normalized equation for a type A distribution surface has the form

$$f_A(\xi_1, \xi_2) = f(\xi_1, \xi_2) +$$

$$+ \sum\sum_{g+h \geq 3} (-1)^{g+h} \frac{c_{gh}}{g! \, h!} \frac{\partial^{g+h} f(\xi_1, \xi_2)}{\partial \xi_1^g \, \partial \xi_2^h}. \quad (146)$$

Here, $f(\xi_1, \xi_2)$ is a normal distribution function of the two random variables, ξ_1 and ξ_2 are the values of these variables, normalized in correspondence with the observed data, and the coefficients c_{gh} are the differences between the standard moments of the given distribution table and the corresponding standard moments of the function $f(\xi_1, \xi_2)$ [cf., (129)].

TABLE 47. Statistics for Table 46

	$r_{h_1|h_2}$				
h_1 \ h_2	0	1	2	3	4
0	1	0	1	−0,698	3,508
1	0	+0,731	−0,485	2,456	
2	1	−0,198	2,101		
3	+0,177	2,077			
4	2,883				

$X_{1(a)} = 32$ cm, $m_{1|0} = +0,311$, $\sigma_1 = 1,854$,

$X_{2(a)} = 25,5$ m, $m_{0|1} = -0,336$, $\sigma_2 = 2,066$,

$$\sqrt{1 - r_{11}^2} = 0,682.$$

TABLE 48. Normalized and Observed Frequencies of Table 46

ξ_1 \ ξ_2	−3,709	−3,225	−2,741	−2,267	−1,773	−1,289	−0,805	−0,321	+0,163	+0,647	+1,131	+1,615	+2,099	Σ
− 2,325	0,2 (1)	0,6 (1)	0,8 (3)	0,6 —	0,5 —	0,4 —	0,3 —	0,2 —	0,1					3,7 (5)
− 1,786	0,2 —	0,8 (1)	2,0 (2)	2,5 (4)	2,8 (6)	3,2 (4)	3,9 (2)	2,4 (1)	0,7 —	0,1 —				18,6 (20)
− 1,247	0,1 —	0,4 —	1,7 (1)	3,8 (2)	5,1 (7)	6,7 (10)	10,2 (12)	9,7 (7)	4,6 (4)	1,1 —	0,1 —			43,5 (43)
− 0,707		0,1 —	0,5 —	1,5 —	5,0 (3)	7,5 (10)	13,7 (14)	20,0 (19)	15,5 (14)	5,8 (7)	1,0 (1)	0,1 (1)		70,7 (69)
− 0,168			0,1 —	0,8 —	2,4 —	4,5 (3)	9,3 (12)	21,0 (21)	26,9 (22)	16,6 (20)	4,7 (8)	0,4 (2)		86,7 (88)
+ 0,372				0,6 —	1,4 (1)	3,0 (3)	10,7 (15)	24,0 (24)	25,1 (20)	11,8 (12)	2,0 (4)			78,6 (79)
+ 0,911						0,4 (2)	2,6 (2)	10,9 (11)	20,1 (21)	16,2 (16)	5,2 (3)	0,1 —		55,5 (55)
+ 1,450							0,1 (1)	2,4 (3)	8,6 (12)	12,0 (8)	7,0 (6)	0,9 (1)		31,0 (31)
+ 1,990								— (1)	0,2 (1)	1,6 (3)	4,9 (4)	5,1 (2)	1,9 (1)	13,7 (12)
+ 2,529										0,5 (1)	1,1 (2)	1,8 (1)	1,4 —	4,8 (4)
+ 3,069											0,1 (1)	0,4 (1)	0,5 —	1,0 (2)
Σ	0,5 (1)	1,9 (2)	5,1 (6)	9,2 (6)	16,4 (16)	23,7 (28)	40,8 (45)	66,7 (67)	85,3 (79)	79,5 (84)	51,9 (52)	22,0 (20)	4,8 (2)	407,8 (408)

In view of the very large standard errors of the moments of order higher than the fourth, it is necessary, in equation (146), to limit ourselves to terms for which

$$g+h \leqslant 4.$$

Using (141), we can put equation (146) in the form

$$f_A(x,\,y)=f(x)\left[f(y)-\frac{r_{013}}{6}f^{(3)}(y)+\frac{r_{014}-3}{24}f^{(4)}(y)\right]+$$

$$+f'(x)\left[-\frac{r_{112}}{2}f''(y)+\frac{r_{113}-3r_{111}}{6}f^{(3)}(y)\right]+$$

$$+ f''(x)\left[-\frac{r_{211}}{2} f'(y) + \frac{r_{212} - 2r_{111}^2 - 1}{4} f''(y)\right] +$$
$$+ f^{(3)}(x)\left[-\frac{r_{310}}{6} f(y) + \frac{r_{311} - 3r_{111}}{6} f'(y)\right] +$$
$$+ f^{(4)} x\left[\frac{r_{410} - 3}{24} f(y)\right]. \tag{147}$$

The transition to normalized frequencies is made via the formula

$$\tilde{n}_{j_1 1/_2} = \frac{n}{\sigma_1\sigma_2 \sqrt{1 - r_{111}^2}} f_A(x, y). \tag{148}$$

As an example, we shall compute the frequencies for the distribution table of diameter, D, and height, H, of pine trees (Table 46). This table was obtained from an enumeration of 180-year-old pine trees not subjected to human ministrations. The necessary statistics for this table are given in Table 47.

By substituting the values of the standard moments in (147), we get

$$f_A(x, y) = f(x)[f(y) + 0,116 f^{(3)}(y) + 0,021 f^{(4)}(y)] +$$
$$+ f'(x)[0,242 f''(y) + 0,044 f^{(3)}(y)] +$$
$$+ f''(x)[0,099 f'(y) + 0,008 f''(y)] +$$
$$+ f^{(3)}(x)[-0,030 f(y) - 0,019 f'(y)] +$$
$$+ f^{(4)}(x)[-0,005 f(y)]. \tag{149}$$

Finally,

$$\tilde{n}_{j_1|j_2} = \frac{408}{1,854 \cdot 2,066 \cdot 0,682} f_A(x, y) = 156,2 f_A(x, y).$$

Computation of the normalized frequencies, using these formulas, is carried out in the following order.

The observed values of the variables in question are replaced by the normalized values

$$\xi_1 = \frac{x_1 - m_{110}}{\sigma_1}, \quad \xi_2 = \frac{x_2 - m_{011}}{\sigma_2}.$$

For a given value

$$x = \xi_1$$

we find the values of the normal probability distribution density function, $f(x)$, and its derivatives from published tables of these numbers.

[Trans. note to the reader: Such a table is found, e.g., in H. Cramér, "Mathematical Methods of Statistics" (Princeton University Press, Princeton, 1946). ESS]

These values remain constant for all cells of the given rows of the distribution table.

Then, for each cell of the distribution table, we compute the value

$$y = \frac{\xi_2 - r_{111}\xi_1}{\sqrt{1 - r_{111}^2}}$$

and find the values of $f(y)$ and its derivatives from the same published tables.

Substituting these values in (149), and performing the necessary calculations, we find $f_A(x, y)$. Multiplying this number by 156,2, we find the frequency sought, $\tilde{n}_{j_1|j_2}$.

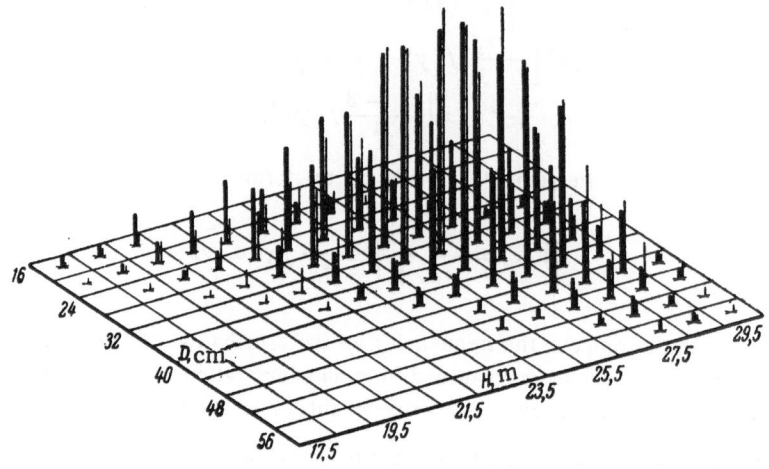

Fig. 10. Observed and normalized frequencies of the distribution table
of diameter and height of 180-year-old pine trees. (Thick lines are observed, thin lines normalized, frequencies.)

We compute, for example, the normalized frequency for the distribution table cell with labels D = 32
cm, H = 24.5 m.

Here,

$$x = -0.168,$$

$$y = \frac{-0.321 - 0.731 \cdot (-0.168)}{\sqrt{1 - (0.731)^2}} = -0.290.$$

From the published tables of $f(x)$ and its derivatives, we find

$$\begin{array}{ll}
f(x) = 0.39335, & f(y) = 0.38251, \\
f'(x) = +0.06608, & f'(y) = +0.11093, \\
f''(x) = -0.38225, & f''(y) = -0.35035, \\
f^{(3)}(x) = -0.19638, & f^{(3)}(y) = -0.32346, \\
f^{(4)}(x) = +1.11371; & f^{(4)}(x) = +0.95723.
\end{array}$$

Consequently,

$$\begin{array}{ll}
f_A(x, y) = 0.39335 [0.38251 - 0.03752 + 0.02010] + & 0.14361 \\
\qquad + 0.06608 [-0.08478 - 0.01423] - & -0.00654 \\
\qquad - 0.38225 [+0.01098 - 0.00280] - & -0.00313 \\
\qquad - 0.19638 [-0.01148 - 0.00211] + & +0.00267 \\
\qquad + 1.11371 [-0.00191] & -0.00213 \\
\hline
& 0.13448.
\end{array}$$

Finally,

$$\tilde{n}_{j_1|j_2} = 156.2 \cdot 0.13448 = 21.0.$$

The results of the computations are given in Table 48 (with observed frequencies given in parentheses),
and are shown graphically on Fig. 10.

MOMENTS

1. Moments of a Distribution

In mathematical statistical investigations, the basic computational work amounts to finding moments. Depending on the value of the random variable with respect to which they are computed, moments are categorized as initial, central, or standard.

An initial moment, m_h, i.e., a moment relative to some initial value, is the sum of the products of each deviation, x_j, raised to a given power, multiplied by the corresponding frequency, n_j, all divided by the sum, n, of all the frequencies:

$$m_h = \frac{\sum_{j=1}^{k} x_{(j)}^h n_j}{n}.$$

(1)

There are two ways of computing initial moments: the method of products and the method of sums.

The computation of moments by the method of products consists of a direct application of the definition of moments (1).

Use of the method of products entails a great deal of computational work, since each frequency of the (discrete) distribution must be multiplied by various powers of the deviation. Moreover, to check the accuracy of the computations, one must either have recourse to special check tables, or repeat the computations which, in the case of a discrete distribution with a large number of items (or groups), presents great difficulty.

In view of this, particular value inheres in a method of computing moments in which finding products is replaced by determining sums and, as a consequence, the computational work reduces to a simple mechanical calculation. With this, not only is the computational labor simplified tremendously but, in the computational process itself, the data are obtained which permit the checking of each step in the work.

The method we have been discussing is called the method of sums.

Basic to the use of the method of sums for computing moments is the existence of completely defined relationships between the sums, s, and differences, d, obtained by this method, on the one hand, and the moments, on the other, these relationships being expressed in the following formulas:

$$\left.\begin{aligned}
m_1 &= \frac{d_1}{n}, \\
m_2 &= \frac{s_1 + 2s_2}{n}, \\
m_3 &= \frac{d_1 + 6d_2 + 6d_3}{n}, \\
m_4 &= \frac{s_1 + 14s_2 + 36s_3 + 24s_4}{n}.
\end{aligned}\right\}$$

(2)

In investigating the distribution of random variables, it ordinarily suffices to limit the calculations to moments of no higher than fourth order.

TABLE 49. Layout for Computing Moments by the Method of Sums

$X_{(j)}$	n_j	(1)	(2)	(3)	(4)	(5)	$x_{(j)}+1$	$(x_j+1)^4 n_j$
48	3	3	3	3	3	—	-3	243
50	17	20	23	26	—	—	-2	272
52	75	95	118	—	—	—	-1	75
54	194	289	—	—	—	—	0	0
56	228						1	228
58	155	274	—	—	—	—	2	2480
60	91	119	152	—	—	—	3	7371
62	23	28	33	38	—	—	4	5888
64	5	5	5	5	5	—	5	3125
—	$n=791$	407	144	29	3	—	Σ	19 682
—	—	426	190	43	5	—		
—	s	833	334	72	8	—		
—	d	$+19$	$+46$	$+14$	$+2$	—		

$m_4^* = \dfrac{19\,682}{791} = 24{,}882$

$$m_1 = \frac{+19}{791} = +0{,}024$$

$$m_2 = \frac{833 + 2\cdot 334}{791} = \frac{1501}{791} = 1{,}898$$

$$m_3 = \frac{19 + 6\cdot 46 + 6\cdot 14}{791} = \frac{+379}{791} = +0{,}479$$

$$m_4 = \frac{833 + 14\cdot 334 + 36\cdot 72 + 24\cdot 8}{791} = \frac{8293}{791} = = 10{,}484$$

$m_0 = 1{,}000$
$4m_1 = 0{,}096$
$6m_2 = 11{,}388$
$4m_3 = 1{,}916$
$m_4 = 10{,}484$

$24{,}884$

Computation of moments by the method of sums requires, first of all, that one construct a table of sums.

Let us calculate, for example, the first four moments of the frequency distribution for the yield strength X (σ_B, kg/mm^2) when stressing axial steel to rupture (Table 6) (see Chapter I), choosing $X_{(a)} = 56$ kg/mm^2. In this case, the table of sums will have the form shown in Table 49.

In addition to a column of variable values and a column of frequencies, the table of sums contains serially numbered columns, one more such column than the number of moments to be computed; in addition, two more columns are placed to the right of the table for checking purposes.

Opposite the frequency corresponding to the initial value $X_{(a)}$ (where $x_j = 0$), one draws a line through all the numbered columns. This has the effect of dividing the table into two parts — an upper and a lower — in each of which summation is performed separately.

In setting up the table of sums, the following procedure is used: for the formation of each number in the table of sums, two numbers are added, one of which is on the same line as, and in the column directly to the left of, the number to be formed, the other being in the same column and directly above (in the upper part of the table) or directly below (in the lower part) the number to be formed. In each portion of the table, these additions are carried out starting from the end of the column and working towards the center of the table, each successive column (from left to right) ending one row sooner than its predecessor.

We note that, with this, even if the frequency of some variable value (or group of values) is zero, this zero frequency is still taken into account in the formation of the number in column (1), i.e., the number in the corresponding row of column (1) is still to be inscribed there.

Beneath each column one writes the sums of the numbers in the upper (negative) and the lower (positive) portions of the column. In addition, under each column one writes the sum, s_h, and the difference, d_h, of these two sums, the difference being obtained by subtracting the upper sum from the lower one.

In setting up the table of sums, one must pay attention to the o r d e r o f c o m p u t a t i o n a l w o r k, this same order being mandatory in the formation of each column of the table of sums:

1. First, one develops the numbers of the column;
2. then, one checks what has been done;
3. finally, one performs the summations of the upper and the lower portions of the column.

By always performing the computational work in the order just stated, one can attain significant speed of computation, along with perfect confidence in the accuracy of what one has done.

Each step in the computations for setting up the table of sums is accompanied by a check.

To verify the correctness of preparation of the first column, one must add the greatest numbers in the upper and lower parts of the first column to the frequency listed opposite the initial value. The sum of these three numbers (which stand immediately above the center line, immediately below it, and directly opposite it) must be equal to the sum of the entire column of frequencies (i.e., equal to n). This assertion is self-evident when one considers the mode of formation of the first column. In our example, we have

$$289 + 274 + 228 = 791$$

Checking the formation of the second and subsequent columns is also very simple: after having written the last number in one portion of the table (i.e., the number closest to the center line of that portion of the given column), one adds to this the last number of the preceding column (in the same portion of the table), and one should obtain the sum of the corresponding portion of the previous column.

Thus, in our example, by adding to the last (i.e., greatest) number, 38, of the positive (i.e., lower) portion of the third column, the last number, 152, of the positive portion of the second column, we obtain the sum of the entire positive portion of the second column, namely, 190.

Once having convinced ourselves that our table of sums has been set up correctly, we turn to the calculation of the moments, using formulas (2). By substituting for s and d in these formulas the corresponding sums and differences from the table of sums, and then performing the indicated operations, we find the moments sought. We inscribe the results of these operations in the lower portion of our layout.

We note that calculation of moments is ordinarily carried out to an accuracy of 0.001.

In order to verify the correctness of our computations of the first four moments, we use the formula

$$m_4^* = m_0 + 4m_1 + 6m_2 + 4m_3 + m_4. \tag{3}$$

In the right member of this formula appear moments with respect to the chosen initial value while, in the left member, is the fourth moment with respect to the value one rank lower than the original one:

$$m_4^* = \frac{\sum_{j=1}^{k} n_j (x_{jp} + 1)^4}{n}. \tag{4}$$

For the calculation of m_4^*, we inscribe, in the penultimate column of our table, the deviation x_j increased by 1, $(x_j + 1)$; then, in the last column of our table we write the product of $(x_j + 1)^4$ by n_j. The sum of these products, divided by the total frequency (n), is m_4^*. The check on the computed moments by formula (3) is entered into the lower right-hand portion of the table.

Initial moments, in themselves, are of little value in applications; they are only auxiliary quantities in the computation of other moments.

Of the initial moments, the only one directly used is the first moment, m_1, from which the mean of the original frequency distribution is calculated.

The formula for computing the mean from the first moment has the form

$$\bar{X} = X_{(a)} + m_1 c.$$

(5)

Thus, to compute the mean, it is necessary to add to the initial value the first moment, multiplied by the interval between values. For example, in our distribution of the yield strength of stressed axial steel, we have

$$X_{(a)} = 56 \text{ kg/mm}^2, \quad m_1 = 0.024, \quad c = 2 \text{ kg/mm}^2.$$

Consequently,

$$\bar{X} = 56 + 0.024 \cdot 2 = 56.048 \text{ kg/mm}^2.$$

The mean is the center about which all the observed values of the random variable are distributed. This center is a constant for the given frequency distribution. By virtue of this, moments computed with respect to this central value are called central moments of the distribution; it is they which are principally used in investigations.

C e n t r a l   m o m e n t s μ_h are defined by the same formula as initial moments except that, instead of deviations from the initial value, one uses deviations from the mean:

$$\mu_h = \frac{\sum_{j=1}^{k} (x_{(j)} - m_1)^h n_j}{n}.$$

(6)

A direct computation of central moments turns out to be quite laborious. This is due to the mean value, with respect to which the central moments are computed, being ordinarily expressed in a number with several decimal places. Therefore, to compute the central moments, one employs the comparatively easily determined initial moments:

$$\left. \begin{aligned} \mu_0 &= 1, \quad \mu_1 = 0, \quad \mu_2 = m_2 - m_1^2, \\ \mu_3 &= m_3 - 3m_2 m_1 + 2m_1^3, \\ \mu_4 &= m_4 - 4m_3 m_1 + 6m_2 m_1^2 - 3m_1^4. \end{aligned} \right\}$$

(7)

To verify the correctness of the computation of the central moments, one uses the formulas

$$\left. \begin{aligned} \mu_3 &= m_3 - 3\mu_2 m_1 - m_1^3, \\ \mu_4 &= m_4 - 4\mu_3 m_1 - 6\mu_2 m_1^2 - m_1^4. \end{aligned} \right\}$$

(8)

For example, the central moments for the frequency distribution of yield strength of stressed axial steel equal

$$\mu_2 = 1.897, \quad \mu_3 = +0.342, \quad \mu_4 = 10.446.$$

The second central moment

$$\mu_2 = m_2 - m_1^2$$

(9)

is called the d i s p e r s i o n (o r   v a r i a n c e) of the distribution, and is denoted by σ^2, while the square root of the dispersion, taken with positive sign, is called the distribution's s t a n d a r d   d e v i a t i o n :

$$\sigma = + \sqrt{\mu_2}.$$
(10)

The standard deviation as computed by this formula, σ, is to be expressed in working units, i.e., it is still an abstract quantity. To make the transition to the units of measurement, one must multiply σ by the interval size, c, in order to obtain the standard deviation, properly so called

$$\bar{\sigma} = \sigma \cdot c.$$
(11)

For example, the standard deviation of the frequency distribution of the yield strength of stressed axial steel equals

$$\sigma = + \sqrt{1.897} = 1.3773, \quad c = 2 \text{ kg/mm}^2, \quad \bar{\sigma} = 2.7546 \text{ kg/mm}^2.$$

By normalizing the deviation x_j in (1),

$$\xi_{(j)} = \frac{x_{(j)} - m_1}{\sigma},$$
(12)

we obtain the standard moments

$$r_h = \frac{\sum\limits_{j=1}^{k} \xi_j^h n_j}{n}.$$
(13)

From (13) we find that

$$r_h = \frac{\mu_h}{\sigma^h},$$

i.e., the standard moments are equal to the ratio of the corresponding central moments to the standard deviation raised to the corresponding power. In particular,

$$r_0 = 1, \quad r_1 = 0, \quad r_2 = 1, \quad r_3 = \frac{\mu_3}{\sigma^3}, \quad r_4 = \frac{\mu_4}{\sigma^4}.$$
(14)

The third standard moment, r_3, serves as a measure of the skewness, α, of the distribution curve:

$$\alpha = r_3,$$
(15)

while the numerical difference between the fourth standard moment and 3 serves as the measure of kurtosis (peakedness), ι, of the distribution curve:

$$\iota = r_4 - 3.$$
(16)

For example, for our frequency distribution of yield strength of stressed axial steel, we have

$$r_3 = +0.131, \quad r_4 = 2.903;$$

so that, for this distribution,

$$\alpha = +0.131, \quad \iota = -0.097.$$

TABLE 50. Layout for Computing $m_{1|1}$ by the Method of Sums

x_1 \ x_2	−4	−3	−2	−1	0	+1	+2	+3	+4	(1) +	(2) −	(3)
−4	—	—	—	—		2	2	2	1	—	7	—7
−3	—	—	—	—		10	7	4	1	—	22	—22
−2	—	—	5	12		69	39	15	2	17	125	— 108
−1	1	5	21	56		146	73	20	2	83	241	—158
0												
+1	2	11	48	147		63	19	4	—	86	208	—122
+2	2	6	28	95		30	12	3	—	45	131	—86
+3	1	3	11	41		9	3	—	—	12	56	—44
+4	—	—	4	12		1	1	—	—	2	16	—14
(1) +	1	5	26	68		103	35	7	—	245	—	—
(2) −	5	20	91	295		227	121	41	6	—	806	—
(3)	−4	—15	— 65	—227		—124	—86	—34	—6	—	—	—561

2. Mixed Moments

An investigation of the relationships between random variables, as with an investigation of a distribution, begins with the computation of the relevant moments. The most important of these moments are the mixed moments of various orders.

The **mixed initial moment** of order (h_1, h_2) of two random variables, X_1 and X_2, is the sum of products over all pairs of deviations, $x_{1(j_1)}$ and $x_{2(j_2)}$, from the corresponding initial values, $X_{1(a)}$ and $X_{2(a)}$, each product consisting of a pair of deviations, each raised to the corresponding power (h_1 or h_2), multiplied by the frequency, $n_{j_1|j_2}$, of the deviation pair, the entire double summation then being divided by the total frequency, n:

$$m_{h_1|h_2} = \frac{\sum_{j_1=1}^{k_1} \sum_{j_2=1}^{k_2} x_{1(j_1)}^{h_1} \, x_{2(j_2)}^{h_2} \, n_{j_1|j_2}}{n}. \tag{17}$$

The first-order mixed moment, $m_{1|1}$, may be computed either by the method of products, or by the method of sums.

The essential disadvantages of the method of products are the greater amount of computational work, and the absence of any possibility of checking the work during the course of computing the mixed moment $m_{1|1}$. For the computation of higher-order mixed moments, this method becomes completely useless.

Therefore, as the practical method of computing mixed moments, one should use the **method of sums**. This method, besides being easier and more convenient, provides the facility of checking the computations at every stage.

Starting with the computation of $m_{1|1}$, we set up, from the tabulated distribution, the first table of sums, the grand total of which, after division by the total frequency (n), is the moment being sought. As an example, we return to our table of joint distribution of toughness and yield strength of stressed axial steel (Table 6).

The structure of the first table of sums (Table 50) is the same as that of the joint distribution table, except that, instead of one total row and one total column, there are now three total rows and three total columns. A null row and a null column, outlined in heavy lines, divide the table of sums into quadrants.

In each of the quadrants of the table of sums, one fills in the numerical entries starting from the extreme corner, and working toward the center. Thus, in the upper left (I) quadrant, the numbers are entered into the table of sums by working toward the center from the left and from above; in the upper right (II) quadrant, from the right and from above; in the lower left (III) quadrant, from the left and from below; in the lower right (IV) quadrant, from the right and from below.

To develop each entry in the table of sums, we use the following general rule:

$$N_{j_1|j_2} = N_{j_1|j_2-1} + N_{j_1-1|j_2} - N_{j_1-1|j_2-1} + n_{j_1|j_2},\qquad(18)$$

i.e., to form each number in the table of sums, one must add two numbers from this table, one of which is in the cell of the same row and the previous column, the other in the cell of the same column and the previous row, then subtract the number in the cell of the previous row and previous column, and then, finally, add the number in the corresponding cell of the joint distribution table (in this case, Table 6).

For example, to form the number 56 in the first quadrant, in the cell at $x_{1(j_1)} = -1$, $x_{2(j_2)} = -1$, we must add the two numbers, 21 and 12, one from the cell in the same row, previous column, the other from the cell in the same column, previous row, and then subtract the number 5 in the cell in the previous row previous column, i.e., in the cell on the same diagonal from the center; finally, to the previous result, we must add the number 28, found in the same (homologous) cell in the joint distribution table (Table 6):

$$21 + 12 - 5 + 28 = 56.$$

Similarly, the number 39 in the second quadrant, in the cell at $x_{1(j_1)} = -2$, $x_{2(j_2)} = +2$, is obtained as the result of the operations

$$15 + 7 - 4 + 21 = 39.$$

To check the correctness of the entries in the table of sums, one must add the largest numbers in each of the four quadrants to the totals of the null row and null column of the joint distribution table, then subtract the frequency of the center cell of this latter table (Table 6, in this case). The result must equal the sum of all the frequencies (n) of the joint distribution table. We have, in our example,

$$56 + 146 + 147 + 63 + 214 + 228 = 854,$$
$$854 - 63 = 791.$$

After this, one finds the sums of the numbers in the rows of the positive (I and IV) quadrants of the table of sums, and writes these in the first total column; then one finds the sums of the numbers in the rows of the

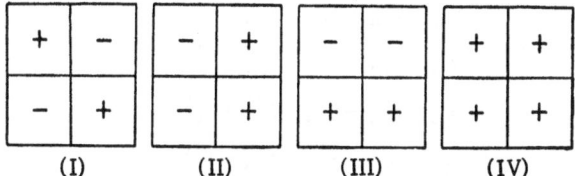

Fig. 11. Diagrams of signs for the computation of
mixed moments by the method of sums.

TABLE 51. Computation of the Mixed Moments $m_{h_1 | 1}$

I				II				III				IV			
—	—	—	—	7	7	7	7	16	16	16	16	2	2	2	2
—	—	—	—	22	29	36	—	56	72	88	—	12	14	16	—
17	17	—	—	125	154	—	—	131	203	—	—	45	59	—	—
83	—	—	—	241	—	—	—	208	—	—	—	86	—	—	—
100	17	—	—	395	190	43	7	411	291	104	16	145	75	18	2

negative (II and III) quadrants, and writes these in the second total column; finally, in the third total column, one inscribes the difference between the numbers in the first and second total columns, subtracting row by row. One then proceeds similarly with the columns of the table of sums. The check on these computations is that the (three) corresponding row and column totals must be equal, i.e., the positive row total must equal the positive column total, etc.

The sum of the numbers in the third total column, as is obvious from its method of formation, equals the numerator of the mixed moment, $m_{1|1}$. By dividing this sum by the sum of all the frequencies in the joint distribution table, we obtain the moment sought:

$$m_{1|1} = \frac{-561}{791} = -0.709,$$

For the computation of the mixed initial moments by the method of sums, one uses the following formulas:

$$m_{1|1} = \frac{d_{1|1}}{n},$$

$$m_{2|1} = \frac{1}{n}(d_{1|1}^{(a)} + 2d_{2|1}^{(a)}),$$

$$m_{1|2} = \frac{1}{n}(d_{1|1}^{(b)} + 2d_{1|2}^{(b)}),$$

$$m_{3|1} = \frac{1}{n}(d_{1|1} + 6d_{2|1} + 6d_{3|1}),$$

$$m_{1|3} = \frac{1}{n}(d_{1|1} + 6d_{1|2} + 6d_{1|3}),$$

$$m_{2|2} = \frac{1}{n}[s_{1|1} + 2(s_{2|1} + s_{1|2}) + 4s_{2|2}].$$

(19)

In these formulas, the numbers d and s are comprised of four components, expressing the corresponding sums of each of the quadrants of the table of sums. To compute the mixed moments, these sums must be taken with the algebraic signs shown in the diagrams of Fig. 11.

Diagram (I) is used in computing mixed moments for which both subscripts are odd, for example, $m_{1|1}$, $m_{3|1}$, and $m_{1|3}$; diagram (II) is used for moments for which the first subscript is even and the second odd, e.g., $m_{2|1}$; diagram (III) is used for moments for which the first subscript is odd and the second even, e.g., $m_{1|2}$; the fourth diagram (IV) is used for moments both of whose subscripts are even, e.g., $m_{2|2}$.

We denote the sum, S, of each quadrant of the table of sums by the corresponding superscript Roman numeral. Then, to compute mixed moments both of whose subscripts are odd, we will have the differences in (19) equal to

$$d_{1|1} = S_{1|1}^{I} - S_{1|1}^{II} - S_{1|1}^{III} + S_{1|1}^{IV}$$

$$d_{2|1} = S_{2|1}^{I} - S_{2|1}^{II} - S_{2|1}^{III} + S_{2|1}^{IV}$$

$$d_{1|2} = S_{1|2}^{I} - S_{1|2}^{II} - S_{1|2}^{III} + S_{1|2}^{IV}$$

$$d_{3|1} = S_{3|1}^{I} - S_{3|1}^{II} - S_{3|1}^{III} + S_{3|1}^{IV}$$

$$d_{1|3} = S_{1|3}^{I} - S_{1|3}^{II} - S_{1|3}^{III} + S_{1|3}^{IV}$$

(20)

For computing moments whose first subscript is even and whose second subscript is odd, the differences will be

$$d_{1|1}^{(a)} = -S_{1|1}^{I} + S_{1|1}^{II} - S_{1|1}^{III} + S_{1|1}^{IV}$$

$$d_{2|1}^{(a)} = -S_{2|1}^{I} + S_{2|1}^{II} - S_{2|1}^{III} + S_{2|1}^{IV} \tag{21}$$

When the first subscript is odd and the second even, the differences for these moments are:

$$d_{1|1}^{(b)} = -S_{1|1}^{I} - S_{1|1}^{II} + S_{1|1}^{III} + S_{1|1}^{IV}$$

$$d_{1|2}^{(b)} = -S_{1|2}^{I} - S_{1|2}^{II} + S_{1|2}^{III} + S_{1|2}^{IV} \tag{22}$$

With both subscripts even, the sums for computation of the moments are

$$s_{1|1} = S_{1|1}^{I} + S_{1|1}^{II} + S_{1|1}^{III} + S_{1|1}^{IV}$$

$$s_{2|1} = S_{2|1}^{I} + S_{2|1}^{II} + S_{2|1}^{III} + S_{2|1}^{IV}$$

$$s_{1|2} = S_{1|2}^{I} + S_{1|2}^{II} + S_{1|2}^{III} + S_{1|2}^{IV}$$

$$s_{2|2} = S_{2|2}^{I} + S_{2|2}^{II} + S_{2|2}^{III} + S_{2|2}^{IV} \tag{23}$$

To compute moments $m_{h_1|1}$, whose first subscript can assume any value, but whose second subscript is unity, we transcribe, from the first table of sums (i.e., Table 50), the row totals, by quadrant, in the first column devoted to each quadrant. By then developing. from these numbers, an ordinary table of sums (cf., the upper portion of Table 49), and then using the corresponding formulas from (19), we find the desired moment(s).

The computations are laid out as shown in Table 51. (The reader will note, in comparing Tables 50 and 51, that the first columns for quadrants I and II in Table 51 duplicate total columns (1) and (2) in the upper part of Table 50, whereas the first columns for quadrants III and IV in Table 51 are total columns (1) and (2) in the lower portion of Table 50, read from left to right, and from bottom to top.)

Here, we have

$$d_{1|1} = 100 - 395 - 411 + 145 = -561,$$
$$d_{2|1} = 17 - 190 - 291 + 75 = -389,$$
$$d_{3|1} = 0 - 43 - 104 + 18 = -129,$$
$$d_{1|1}^{(a)} = -100 + 395 - 411 + 145 = +29,$$
$$d_{2|1}^{(a)} = -17 + 190 - 291 + 75 = -43.$$

Consequently,

$$m_{1|1} = \frac{-561}{791} = -0.709,$$

$$m_{2|1} = \frac{29 + 2 \cdot (-43)}{791} = -0.072,$$

$$m_{3|1} = \frac{-561 + 6 \cdot (-389) + 6 \cdot (-129)}{791} = -4.638.$$

Similarly, to compute mixed moments $m_{1|h_2}$, whose first subscript is unity and whose second subscript is variable, we transcribe, from the first table of sums (again, Table 50), the column totals, by quadrant, in what we then develop into an ordinary table of sums. Substituting the numbers thus obtained into the corresponding formulas of (19), we find the moment(s) sought.

TABLE 52. Computation of the Mixed Moments $m_1|h_2$

I				II				III				IV			
1	1	1	1	6	6	6	6	5	5	5	5	—	—	—	—
5	6	7	—	41	47	53	—	20	25	30	—	7	7	7	—
26	32	—	—	121	168	—	—	91	116	—	—	35	42	—	—
68	—	—	—	227	—	—	—	295	—	—	—	103	—	—	—
100	39	8	1	395	221	59	6	411	146	35	5	145	49	7	—

The computations are laid out as in Table 52. (The reader is again urged to compare the order and disposition of the total rows in Table 50 with the first columns of the quadrant layouts in Table 52.)

Here, we have

$$d_{1|1} = 100 - 395 - 411 + 145 = -561,$$
$$d_{1|2} = 39 - 221 - 146 + 49 = -279,$$
$$d_{1|3} = 8 - 59 - 35 + 7 = -79,$$
$$d_{111}^{(b)} = -100 - 395 + 411 + 145 = +61,$$
$$d_{112}^{(b)} = -39 - 221 + 146 + 49 = -65.$$

Consequently,

$$m_{1|1} = \frac{-561}{791} = -0.709,$$
$$m_{1|2} = \frac{61 + 2 \cdot (-65)}{791} = -0.087,$$
$$m_{1|3} = \frac{-561 + 6 \cdot (-279) + 6 \cdot (-79)}{791} = -3.425.$$

TABLE 53. Layout for Computing $m_2|_2$ by the Method of Sums

x_1 \ x_2	-4	-3	-2	-1	0	$+1$	$+2$	$+3$	$+4$	Σ
-4				—		—	5	3	1	9
-3				—		—	17	8	2	27
-2			5	—		—	73	25	4	107
-1	—	—	—	—		—	—	—	—	—
0										
$+1$	—	—	—	—		—	—	—	—	—
$+2$	3	12	55	—		—	19	3		92
$+3$	1	4	19	—		—	4			28
$+4$			4	—		—	1			5
Σ	4	16	83	—		—	119	39	7	268

To compute $m_{2|2}$, we construct a s e c o n d t a b l e o f s u m s, constructed from the first table of sums precisely as this table was constructed from the original joint distribution table, with those columns and rows neighboring the null column and row in the first table of sums left blank in the second (Table 53).

In order to check the correctness of the entries in the second table of sums, one adds to the largest number in each of the quadrants of this table, i.e., to the numbers 5, 73, 55, and 19 of Table 53, all the numbers from the first table of sums (Table 50) which correspond to deviations of $+1$ or -1 (i.e., one adds the numbers in each quadrant which border on the null row and column). The total thus obtained must equal the total of all the numbers in the first table of sums, i.e.,

$$245 + 806 = 1051.$$

In our example, we have

$$5 + 73 + 55 + 19 + 95 + 322 + 356 + 126 = 1051.$$

After having verified the correctness of our layout of the second table of sums, we write the totals of each row and column of this table, and then find their sum. This sum is precisely $s_{2|2}$. In the given case,

$$s_{2|2} = 268.$$

We also find the other sums entering into formulas (19) for moment $m_{2|2}$, using expression (23) for this purpose. We then have (cf., Tables 51 and 52)

$$s_{1|1} = 100 + 395 + 145 + 411 = 1051,$$
$$s_{2|1} = 17 + 190 + 291 + 75 = 573,$$
$$s_{1|2} = 39 + 221 + 146 + 49 = 455.$$

Thus,

$$m_{2|2} = \frac{1}{791} \left[1051 + 2 \cdot (573 + 455) + 4 \cdot 268 \right] = \frac{4179}{791} = 5.283.$$

On the basis of the mixed initial moments, we find the m i x e d c e n t r a l m o m e n t s from the formulas

$$\left.\begin{aligned}
\mu_{1|1} &= m_{1|1} - m_{1|0} m_{0|1}, \\
\mu_{2|1} &= m_{2|1} - 2 m_{1|1} m_{1|0} - m_{0|1}(\mu_{2|0} - m_{1|0}^2), \\
\mu_{1|2} &= m_{1|2} - 2 m_{1|1} m_{0|1} - m_{1|0}(\mu_{0|2} - m_{0|1}^2), \\
\mu_{3|1} &= m_{3|1} - 3 m_{2|1} m_{1|0} + 3 m_{1|1} m_{1|0}^2 - m_{0|1}(\mu_{3|0} - m_{1|0}^3), \\
\mu_{1|3} &= m_{1|3} - 3 m_{1|2} m_{0|1} + 3 m_{1|1} m_{0|1}^2 - m_{1|0}(\mu_{0|3} - m_{0|1}^3), \\
\mu_{2|2} &= m_{2|2} - 2 (m_{2|1} m_{0|1} + m_{1|2} m_{1|0}) + 4 m_{1|1} m_{1|0} m_{0|1} + m_{1|0}^2(\mu_{0|2} - m_{0|1}^2) + m_{0|1}^2 \mu_{2|0}.
\end{aligned}\right\} \quad (24)$$

We check our computations by means of the formulas

$$\left.\begin{aligned}
\mu_{1|1} &= m_{1|1} - m_{1|0} m_{0|1}, \\
\mu_{2|1} &= m_{2|1} - 2 \mu_{1|1} m_{1|0} - m_{2|0} m_{0|1}, \\
\mu_{1|2} &= m_{1|2} - 2 \mu_{1|1} m_{0|1} - m_{0|2} m_{1|0}, \\
\mu_{3|1} &= m_{3|1} - 3 \mu_{2|1} m_{1|0} - 3 \mu_{1|1} m_{1|0}^2 - m_{3|0} m_{0|1}, \\
\mu_{1|3} &= m_{1|3} - 3 \mu_{1|2} m_{0|1} - 3 \mu_{1|1} m_{0|1}^2 - m_{0|3} m_{1|0}, \\
\mu_{2|2} &= m_{2|2} - 2 (\mu_{2|1} m_{0|1} + \mu_{1|2} m_{1|0}) - 4 \mu_{1|1} m_{1|0} m_{0|1} - m_{0|2} m_{1|0}^2 - \mu_{2|0} m_{0|1}^2.
\end{aligned}\right\} \quad (25)$$

The mixed standard moments are found from the central moments:

$$r_{h_1|h_2} = \frac{\mu_{h_1|h_2}}{\sigma_1^{h_1} \sigma_2^{h_2}}. \tag{26}$$

In particular,

$$\left.\begin{array}{lll} r_{1|1} = \dfrac{\mu_{1|1}}{\sigma_1 \sigma_2}, & r_{2|1} = \dfrac{\mu_{2|1}}{\sigma_1^2 \sigma_2}, & r_{1|2} = \dfrac{\mu_{1|2}}{\sigma_1 \sigma_2^2}, \\[2ex] r_{3|1} = \dfrac{\mu_{3|1}}{\sigma_1^3 \sigma_2}, & r_{1|3} = \dfrac{\mu_{1|3}}{\sigma_1 \sigma_2^3}, & r_{2|2} = \dfrac{\mu_{2|2}}{\sigma_1^2 \sigma_2^2}. \end{array}\right\} \tag{27}$$

The first-order mixed standard moment, $r_{1|1}$, is called the correlation coefficient, and is symbolized by r:

$$r = \frac{\mu_{1|1}}{\sigma_1 \sigma_2}. \tag{28}$$

By using formulas (24), (25), and (26), we can now find mixed central and standard moments for the joint distribution of toughness and yield strength of stressed axial steel (Table 6):

$$\begin{array}{ll} \mu_{1|1} = -0.714, & r_{1|1} = -0,335, \\ \mu_{2|1} = 0.176, & r_{2|1} = 0,053, \\ \mu_{1|2} = -0.332, & r_{1|2} = -0.113, \\ \mu_{3|1} = -4.717, & r_{3|1} = -0,920, \\ \mu_{1|3} = -3.495, & r_{1|3} = -0,889, \\ \mu_{2|2} = 5.422, & r_{2|2} = 1,201. \end{array}$$

3. Conditional Moments

In addition to the moments considered above, known as total moments, valuable moments for investigative purposes are the so-called conditional moments.

The conditional moment, $m_{(j_1)|h_2}$ of order h_2 of variable X_2, given that the observed value of variable X_1 fell in the class $X_{1(j_1)}$, is defined as the sum of products of the given power of the deviations, $x_{2(j_2)}^{h_2}$, by the corresponding frequency $p_{(j)|j_2} = n_{j_1|j_2}/n_{j_1|\cdot}$:

$$m_{(j_1)|h_2} = \sum_{j_2=1}^{k_2} p'_{(j_1)|j_2} x_{2(j_2)}^{h_2}. \tag{28}$$

In particular,

$$m_{(j_1)|1} = \sum_{j_2=1}^{k_2} p'_{(j_1)|j_2} x_{2(j_2)}, \tag{29}$$

$$m_{(j_1)|2} = \sum_{j_2=1}^{k_2} p'_{(j_1)|j_2} x_{2(j_2)}^2. \tag{30}$$

By taking (17) and (28) into account, we find that

$$m_{h_1|h_2} = \sum_{j_1=1}^{k_1} p'_{j_1|\cdot} x_{1(j_1)}^{h_1} m_{(j_1)|h_2} \tag{31}$$

(the theorem of multiplication of moments), where $p'_{j_1|\cdot} = \dfrac{n_{j_1|\cdot}}{n}$. In particular,

$$m_{0|1} = \sum_{j_1=1}^{k_1} p_{j_1|\cdot} \, m_{(j_1)|1}, \tag{32}$$

$$m_{1|1} = \sum_{j_1=1}^{k_1} p'_{j_1|\cdot} \, x_{1(j_1)} m_{(j_1)|1}. \tag{33}$$

Conditional central moments have the form

$$\mu^*_{(j_1)|h_2} = \sum_{j_2=1}^{k_2} p'_{(j_1)|j_2} (x_{2(j_2)} - m_{(j_1)|1})^{h_2} \tag{34}$$

and

$$\mu_{(j_1)|h_2} = \sum_{j_2=1}^{k_2} p_{(j_1)|j_2} (x_{2(j_2)} - m_{0|1})^{h_2}. \tag{35}$$

In particular,

$$\left.\begin{aligned} \mu^*_{(j_1)|1} &= 0, \\ \mu^*_{(j_1)|2} &= m_{(j_1)|2} - m^2_{(j_1)|1}; \end{aligned}\right\} \tag{36}$$

$$\left.\begin{aligned} \mu_{(j_1)|1} &= m_{(j_1)|1} - m_{0|1}, \\ \mu_{(j_1)|2} &= \mu^*_{(j_1)|2} + (m_{(j_1)|2} - m_{0|1})^2. \end{aligned}\right\} \tag{37}$$

In the case of central moments, formula (31) assumes the form

$$\mu_{h_1|h_2} = \sum_{j_1=1}^{k_1} p'_{j_1|\cdot} (x_{1(j_1)} - m_{1|0})^{h_1} \mu_{(j_1)|h_2}. \tag{38}$$

In particular, taking (37) into account, we get

$$\mu_{1|1} = \sum_{j_1=1}^{k_1} p'_{j_1|\cdot} (x_{1(j_1)} - m_{1|0})(m_{(j_1)|1} - m_{0|1}); \tag{39}$$

$$\mu_{0|2} = \sum_{j_1=1}^{k_1} p'_{j_1|\cdot} \, \mu_{(j_1)|2}, \tag{40}$$

or

$$\mu_{0|2} = \sum_{j_1=1}^{k_1} p'_{j_1|\cdot} \, \mu^*_{(j_1)|2} + \sum_{j_1=1}^{k_1} p'_{j_1|\cdot} (m_{(j_1)|1} - m_{0|1})^2. \tag{41}$$

Thus, the total dispersion, $\mu_{0|2}$, of the variable X_2 equals the mean of the conditional dispersion, $\mu_{(j_1)|2}$, added to the dispersion of the conditional means of the values of $m_{(j_1)|1}$ about the total mean value, $m_{0|1}$.

TABLE 54. Computation of Conditional Moments

$X_{(j)}$	n_j	(1)	(2)	(3)	$x_{(j)}+1$	$(x_{(j)}+1)^2 n_j$
52	5	5	5	—	−1	5
54	7	12	—	—	0	0
56	21				1	21
58	27	59	—	—	2	108
60	21	32	44	—	3	189
62	10	11	12	13	4	160
64	1	1	1	1	5	25
—	92	17	5	—	Σ	508
—	—	103	57	—		
—	s	120	62	—	$m^*_{(j_1)|2} =$	
—	d	+86	—	—	$= \dfrac{508}{92} = \mathbf{5{,}522}$	

$$m_{(j_1)|1} = \frac{+86}{92} = +0{,}935$$

$$m_{(j_1)|2} = \frac{120 + 2 \cdot 62}{92} = 2{,}652$$

$$\mu^*_{(j_1)|2} = 2{,}652 - 0{,}935^2 = 1{,}778$$

$$\bar{X}_{(j_1)|1} = 56 + 0{,}935 \cdot 2 = 57{,}87$$

$$m_{(j_1)|0} = 1$$

$$2m_{(j_1)|1} = 1{,}870$$

$$\underline{m_{(j_1)|2} = 2{,}652}$$

$$\mathbf{5{,}522}$$

TABLE 55. Conditional Moments for Each Row of Table 6

$X_{1(j_1)}$	$n_{j_1|\cdot}$	$m_{(j_1)|1}$	$m_{(j_1)|2}$	$\mu^*_{(j_1)|2}$	$\bar{X}_{(j_1)|1}$
4,5	2	+ 3,500	12,500	0,250	63,00
5,5	10	+ 1,500	3,300	1,050	59,00
6,5	92	+ 0,935	2,652	1,778	57,87
7,5	169	+ 0,296	1,944	1,853	56,59
8,5	214	− 0,079	1,734	1,727	55,84
9,5	132	− 0,273	1,485	1,410	55,45
10,5	103	− 0,408	1,845	1,678	55,18
11,5	53	− 0,566	1,623	1,302	54,87
12,5	16	− 0,875	1,750	0,984	54,25
—	791	+ 0,024	1,898	1,897	56,05

TABLE 56. Computation of the Correlation Coefficient by Means of Conditional Initial Moments

$X_{1(j_1)}$	$n_{j_1|\cdot}$	$x_{1(j_1)}$	$m_{(j_1)|1}$	$n_{j_1|\cdot} x_{1(j_1)} m_{(j_1)|1}$
4,5	2	− 4	+ 3,500	− 28,000
5,5	10	− 3	+ 1,500	− 45,000
6,5	92	− 2	+ 0,935	− 172,040
7,5	169	− 1	+ 0,296	− 50,024
8,5	214	0	− 0,072	0
9,5	132	1	− 0,273	− 36,036
10,5	103	2	− 0,408	− 84,048
11,5	53	3	− 0,566	− 89,994
12,5	16	4	− 0,875	− 56,000
Σ	791	—	—	− 561,142

$$m_{111} = \frac{-561{,}142}{791} = -0{,}709,$$

$$r = \frac{-0{,}709 - 0{,}215 \cdot 0{,}024}{1{,}550 \cdot 1{,}377} = \frac{-0{,}714}{2{,}134} = -0{,}335.$$

TABLE 57. Layout for Computing
the Correlation Ratio by Formula (50)

$X_{1(j_1)}$	$n_{j_1|\cdot}$	$m_{(j_1)|1}$	$m_{(j_1)|1}^2$	$m_{(j_1)|1}^2 n_{j_1|\cdot}$
4,5	2	$+3,500$	12,2500	24,5000
5,5	10	$+1,500$	2,2500	22,5000
6,5	92	$+0,935$	0,8742	80,4264
7,5	169	$+0,296$	0,0876	14,8044
8,5	214	$-0,079$	0,0062	1,3268
9,5	132	$-0,273$	0,0745	9,8340
10,5	103	$-0,408$	0,1665	17,1495
11,5	53	$-0,566$	0,3204	16,9812
12,5	16	$-0,875$	0,7656	12,2496
Σ	791	—	—	199,7719

$$\eta_{21}^2 = \frac{1}{1,897}\left(\frac{199,7719}{791} - 0,024^2\right) =$$
$$= \frac{0,252}{1,897} = 0,1328,$$
$$\eta_{21} = \sqrt{0,1328} = 0,3644.$$

TABLE 58. Layout for
Computing the Correlation Ratio
by Formula (51)

$X_{1(j_1)}$	$n_{j_1|\cdot}$	$\mu_{(j_1)|2}^*$	$\mu_{(j_1)|2}^* n_{j_1|\cdot}$
4,5	2	0,250	0,500
5,5	10	1,050	10,500
6,5	92	1,778	163,576
7,5	169	1,8,3	313,157
8,5	214	1,727	369,578
9,5	132	1,410	186,120
10,5	103	1,678	172,834
11,5	53	1,302	69,006
12,5	16	0,984	15,744
Σ	791	—	1301,015

$$\frac{1301,015}{791} = 1,6451,$$
$$\frac{1,645}{1,897} = 0,8672,$$
$$1 - 0,8672 = 0,1328,$$
$$\sqrt{0,1328} = 0,3644 = \eta_{21}.$$

The conditional standard moments have the form

$$r_{(j_1)|h_2}^* = \frac{\mu_{(j_1)|h_2}^*}{\mu_{(j_1)|2}^{\frac{1}{2}h_2}}, \tag{42}$$

$$r_{(j_1)|h_2} = \frac{\mu_{(j_1)|h_2}}{\mu_{0|2}^{\frac{1}{2}h_2}}. \tag{43}$$

In particular,

$$r_{(j_1)|1} = \frac{m_{(j_1)|1} - m_{0|1}}{\mu_{0|2}^{\frac{1}{2}}}. \tag{44}$$

For standard moments, formula (31) takes the form

$$r_{h_1|h_2} = \sum_{j_1=1}^{k_1} p'_{j_1|\cdot} \, \xi_{1(j_1)}^{h_1} r_{(j_1)|h_2}. \tag{45}$$

In particular,

$$r_{1|1} = \sum_{j_1=1}^{k_1} p'_{j_1|\cdot} \xi_{1(j_1)} r_{(j_1)|1}. \tag{46}$$

From (44), we find

$$m_{(j_1)|1} = r_{(j_1)|1}\sigma_2 + m_{0|1}.$$

Consequently, the conditional mean of the variable X_2 will, according to (5), equal

$$\overline{X}_{(j_1)|1} = \overline{X}_{2(a)} + (r_{(j_1)|1}\sigma_2 + m_{0|1})c_2 = \overline{X}_2 + r_{(j_1)|1}\sigma_2. \tag{47}$$

The conditional standard first-order moment is used in constructing a statistic characterizing the relationship between random variables, this statistic being called the correlation ratio.

The correlation ratio of X_2 to X_1 equals the square root of the mean of the squares of the conditional standard first-order moments, $r_{(j_1)|1}$, and is symbolized by η_{21}:

$$\eta_{21}^2 = \sum_{j_1=1}^{k_1} p'_{j_1|\cdot} r^2_{(j_1)|1}. \tag{48}$$

Similarly, the correlation ratio of X_1 to X_2 equals

$$\eta_{12}^2 = \sum_{j_2=1}^{k_2} p'_{\cdot|j_2} r^2_{1|(j_2)}. \tag{49}$$

Taking (48), (44), and (32) into account, we obtain the formulas for computing the correlation ratio:

$$\eta_{21}^2 = \frac{\sum\limits_{j_1=1}^{k_1} p'_{j_1|\cdot} (m_{(j_1)|1} - m_{0|1})^2}{\mu_{0|2}} = \frac{\sum\limits_{j_1=1}^{k_1} p'_{j_1|\cdot} m^2_{(j_1)|1} - m^2_{0|1},}{\mu_{0|2}}, \tag{50}$$

$$\eta_{21}^2 = 1 - \frac{\sum\limits_{j_1=1}^{k_1} p'_{j_1|\cdot} \mu^*_{(j_1)|2}}{\mu_{0|2}} \tag{51}$$

and the analogous formulas for η_{12}^2.

We should also note the following. If the number of joint observations of the random variables X_1 and X_2 is small (and, consequently, the joint distribution table lacks reliability), then there will ordinarily be a one-to-one correspondence between the values of the two variables. In this case,

$$\sum_{j_1=1}^{k_1} p_{j_1|\cdot} r^2_{(j_1)|1} = \sum_{j_2=1}^{k_2} p'_{\cdot|j_2} r^2_{1|(j_2)} = 1. \tag{52}$$

Conversely, if (52) holds, then such a correspondence always exists.

Computation of the conditional moments for each row and column of the joint distribution table, i.e., the moments of the nominal classes of the distribution, is performed by the same methods as the computation of the moments of the total classes of the distribution. For this, it is only necessary to retain the in v a r i a n c e of the initial value in all computations of moments.

Let us calculate, for example, the conditional moments for each row of the joint distribution table of toughness and yield strength for stressed axial steel (Table 6), retaining invariable the initial value adopted for the distribution of yield strength, namely, $X_{2(a)} = 56$ kg/mm^2 (cf., Table 49).

For the row labeled $a_k = 6.5$ kgm/cm^2, the computation of the conditional moments of the distribution of yield strength, X_2 (σ_B in kg/mm^2), is laid out in Table 54. The conditional moments for each row of Table 6 are given in Table 55.

Based on the conditional moments we have found, we may calculate the correlation coefficient, r, between toughness and yield strength of stressed axial steel, as well as the correlation ratio, η_{21}, of the yield strength to toughness of the stressed axial steel.

By using formula (33), we can calculate the correlation coefficient, as laid out in Table 56.

The computation of the correlation ratio η_{21}, from the first conditional moments, $m_{(j_1)|1}$, using formula (50), is shown on Table 57. The check on the computation of the correlation ratio η_{21} is performed by means of formula (51) (Table 58).

TABLE I. Powers of Integers

n	n²	n³	n⁴	n⁵	n⁶	n⁷
1	1	1	1	1	1	1
2	4	8	16	32	64	128
3	9	27	81	243	729	2 187
4	16	64	256	1 024	4 096	16 384
5	25	125	625	3 125	15 625	78 125
6	36	216	1 296	7 776	46 656	279 936
7	49	343	2 401	16 807	117 649	823 543
8	64	512	4 096	32 768	262 144	2 097 152
9	81	729	6 561	59 049	531 441	4 782 969
10	100	1 000	10 000	100 000	1 000 000	10 000 000
11	121	1 331	14 641	161 051	1 771 561	19 487 171
12	144	1 728	20 736	248 832	2 985 984	35 831 808
13	169	2 197	28 561	371 293	4 826 809	62 748 517
14	196	2 744	38 416	537 824	7 529 536	105 413 504
15	225	3 375	50 625	759 375	11 390 625	170 859 375
16	256	4 096	65 536	1 048 576	16 777 216	268 435 456
17	289	4 913	83 521	1 419 857	24 137 569	410 338 673
18	324	5 832	104 976	1 889 568	34 012 224	612 220 032
19	361	6 859	130 321	2 476 099	47 045 881	893 871 739
20	400	8 000	160 000	3 200 000	64 000 000	1 280 000 000
21	441	9 261	194 481	4 084 101	85 766 121	1 801 088 541
22	484	10 648	234 256	5 153 632	113 379 904	2 494 357 888
23	529	12 167	279 841	6 436 343	148 035 889	3 404 825 447
24	576	13 824	331 776	7 962 624	191 102 976	4 586 471 424
25	625	15 625	390 625	9 765 625	244 140 625	6 103 515 625
26	676	17 576	456 976	11 881 376	308 915 776	8 031 810 176
27	729	19 683	531 441	14 348 907	387 420 489	10 460 353 203
28	784	21 952	614 656	17 210 368	481 890 304	13 492 928 512
29	841	24 389	707 281	20 511 149	594 823 321	17 249 876 309
30	900	27 000	810 000	24 300 000	729 000 000	21 870 000 000
31	961	29 791	923 521	28 629 151	887 503 681	27 512 614 111
32	1 024	32 768	1 048 576	33 554 432	1 073 741 824	34 359 738 368
33	1 089	35 937	1 185 921	39 135 393	1 291 467 969	42 618 442 977
34	1 156	39 304	1 336 336	45 435 424	1 544 804 416	52 523 350 144
35	1 225	42 875	1 500 625	52 521 875	1 838 265 625	64 339 296 875
36	1 296	46 656	1 679 616	60 466 176	2 176 782 336	78 364 164 096
37	1 369	50 653	1 874 161	69 343 957	2 565 726 409	94 931 877 133
38	1 444	54 872	2 085 136	79 235 168	3 010 936 384	114 415 582 592
39	1 521	59 319	2 313 441	90 224 199	3 518 743 761	137 231 006 679
40	1 600	64 000	2 560 000	102 400 000	4 096 000 000	163 840 000 000
41	1 681	68 921	2 825 761	115 856 201	4 750 104 241	194 754 273 881
42	1 764	74 088	3 111 696	130 691 232	5 489 031 744	230 539 333 248
43	1 849	79 507	3 418 801	147 008 443	6 321 363 049	271 818 611 107
44	1 936	85 184	3 748 096	164 916 224	7 256 313 856	319 277 809 664
45	2 025	91 125	4 100 625	184 528 125	8 303 765 625	373 669 453 125
46	2 116	97 336	4 477 456	205 962 976	9 474 296 896	435 817 657 216
47	2 209	103 823	4 879 681	229 345 007	10 779 215 329	506 623 120 463
48	2 304	110 592	5 308 416	254 803 968	12 230 590 464	587 068 342 272
49	2 401	117 649	5 764 801	282 475 249	13 841 287 201	678 223 072 849
50	2 500	125 000	6 250 000	312 500 000	15 625 000 000	781 250 000 000

n	n²	n³	n⁴	n⁵	n⁶	n⁷
51	2 601	132 651	6 765 201	345 025 251	17 596 287 801	897 410 677 851
52	2 704	140 608	7 311 616	380 204 032	19 770 609 664	1 028 071 702 528
53	2 809	148 877	7 890 481	418 195 493	22 164 361 129	1 174 711 139 837
54	2 916	157 464	8 503 056	459 165 024	24 794 911 296	1 338 925 209 984
55	3 025	166 375	9 150 625	503 284 375	27 680 640 625	1 522 435 234 375
56	3 136	175 616	9 834 496	550 731 776	30 840 979 456	1 727 094 849 536
57	3 249	185 193	10 556 001	601 692 057	34 296 447 249	1 954 897 493 193
58	3 364	195 112	11 316 496	656 356 768	38 068 692 544	2 207 984 167 552
59	3 481	205 379	12 117 361	714 924 299	42 180 533 641	2 488 651 484 819
60	3 600	216 000	12 960 000	777 600 000	46 656 000 000	2 799 360 000 000
61	3 721	226 981	13 845 841	844 596 301	51 520 374 361	3 142 742 836 021
62	3 844	238 328	14 776 336	916 132 832	56 800 235 584	3 521 614 606 208
63	3 969	250 047	15 752 961	992 436 543	62 523 502 209	3 938 980 639 167
64	4 096	262 144	16 777 216	1 073 741 824	68 719 476 736	4 398 046 511 104
65	4 225	274 625	17 850 625	1 160 290 625	75 418 890 625	4 902 227 890 625
66	4 356	287 496	18 974 736	1 252 332 576	82 653 950 016	5 455 160 701 056
67	4 489	300 763	20 151 121	1 350 125 107	90 458 382 169	6 060 711 605 323
68	4 624	314 432	21 381 376	1 453 933 568	98 867 482 624	6 722 988 818 432
69	4 761	328 509	22 667 121	1 564 031 349	107 918 163 081	7 446 353 252 589
70	4 900	343 000	24 010 000	1 680 700 000	117 649 000 000	8 235 430 000 000
71	5 041	357 911	25 411 681	1 804 229 351	128 100 283 921	9 095 120 158 391
72	5 184	373 248	26 873 856	1 934 917 632	139 314 069 504	10 030 613 004 288
73	5 329	389 017	28 398 241	2 073 071 593	151 334 226 289	11 047 398 519 097
74	5 476	405 224	29 986 576	2 219 006 624	164 206 490 176	12 151 206 273 024
75	5 625	421 875	31 640 625	2 373 046 875	177 978 515 625	13 348 388 671 875
76	5 776	438 976	33 362 176	2 535 525 376	192 699 928 576	14 645 194 571 776
77	5 929	456 533	35 153 041	2 706 784 157	208 422 380 089	16 048 523 266 853
78	6 084	474 552	37 015 056	2 887 174 368	225 199 600 704	17 565 568 854 912
79	6 241	493 039	38 950 081	3 077 056 399	243 087 455 521	19 203 908 986 159
80	6 400	512 000	40 960 000	3 276 800 000	262 144 000 000	20 971 520 000 000
81	6 561	531 441	43 046 721	3 486 784 401	282 429 536 481	22 876 792 454 961
82	6 724	551 368	45 212 176	3 707 398 432	304 006 671 424	24 928 547 056 763
83	6 889	571 787	47 458 321	3 939 040 643	326 940 373 369	27 136 050 989 627
84	7 056	592 704	49 787 136	4 182 119 424	351 298 031 616	29 509 034 655 744
85	7 225	614 125	52 200 625	4 437 053 125	377 149 515 625	32 057 708 828 125
86	7 396	636 056	54 700 816	4 704 270 176	404 567 235 136	34 792 782 221 696
87	7 569	658 503	57 289 761	4 984 209 207	433 626 201 009	37 725 479 487 783
88	7 744	681 472	59 969 536	5 277 319 168	464 404 086 784	40 867 559 636 992
89	7 921	704 969	62 742 241	5 584 059 449	496 981 290 961	44 231 334 895 529
90	8 100	729 000	65 610 000	5 904 900 000	531 441 000 000	47 829 690 000 000
91	8 281	753 571	68 574 961	6 240 321 451	567 869 252 041	51 676 101 935 731
92	8 464	778 688	71 639 296	6 590 815 232	606 355 001 344	55 784 660 123 648
93	8 649	804 357	74 805 201	6 956 883 693	646 990 183 449	60 170 087 060 757
94	8 836	830 584	78 074 896	7 339 040 224	689 869 781 056	64 847 759 419 264
95	9 025	857 375	81 450 625	7 737 809 375	735 091 890 625	69 833 729 609 375
96	9 216	884 736	84 934 656	8 153 726 976	782 757 789 696	75 144 747 810 816
97	9 409	912 673	88 529 281	8 587 340 257	832 972 004 929	80 798 284 478 113
98	9 604	941 192	92 236 816	9 039 207 968	885 842 380 864	86 812 553 324 672
99	9 801	970 299	96 059 601	9 509 900 499	941 480 149 401	93 206 534 790 699
100	10 000	1 000 000	100 000 000	10 000 000 000	1 000 000 000 000	100 000 000 000 000

TABLE II. Chebyshev Numbers from n = 3 to n = 50

3		4			5				6				
ψ_1	$3\psi_2$	$2\psi_1$	ψ_2	$\frac{10}{3}\psi_3$	ψ_1	ψ_2	$\frac{5}{6}\psi_3$	$\frac{35}{12}\psi_4$	$2\psi_1$	$\frac{3}{2}\psi_2$	$\frac{5}{3}\psi_3$	$\frac{7}{12}\psi_4$	$\frac{21}{10}\psi_5$
−1	+1	−3	+1	−1	−2	+2	−1	+1	−5	+5	−5	+1	− 1
0	−2	−1	−1	+3	−1	−1	+2	−4	−3	−1	+7	−3	+ 5
+1	+1	+1	−1	−3	0	−2	0	+6	−1	−4	+4	+2	−10
		+3	+1	+1	+1	−1	−2	−4	+1	−4	−4	+2	+10
					+2	+2	+1	+1	+3	−1	−7	−3	− 5
									+5	+5	+5	+1	+ 1
2	6	20	4	20	10	14	10	70	70	84	180	28	252

7					8					9				
ψ_1	ψ_2	$\frac{1}{6}\psi_3$	$\frac{7}{12}\psi_4$	$\frac{7}{20}\psi_5$	$2\psi_1$	ψ_2	$\frac{2}{3}\psi_3$	$\frac{7}{12}\psi_4$	$\frac{7}{10}\psi_5$	ψ_1	$3\psi_2$	$\frac{5}{6}\psi_3$	$\frac{7}{12}\psi_4$	$\frac{3}{20}\psi_5$
−3	+5	−1	+3	−1	−7	+7	−7	+ 7	− 7	−4	+28	−14	+14	− 4
−2	0	+1	−7	+4	−5	+1	+5	−13	+23	−3	+ 7	+ 7	−21	+11
−1	−3	+1	+1	−5	−3	−3	+7	− 3	−17	−2	− 8	+13	−11	− 4
0	−4	0	+6	0	−1	−5	+3	+ 9	−15	−1	−17	+9	+ 9	− 9
+1	−3	−1	+1	+5	+1	−5	−3	+ 9	+15	0	−20	0	+18	0
+2	0	−1	−7	−4	+3	−3	−7	− 3	+17	+1	−17	−9	+ 9	+ 9
+3	+5	+1	+3	+1	+5	+1	−5	−13	−23	+2	− 8	−13	−11	+ 4
					+7	+7	+7	+ 7	+ 7	+3	+ 7	− 7	−21	−11
										+4	+28	+14	+14	+ 4
28	84	6	154	84	168	168	264	616	2184	60	2772	990	2002	468

10					11					12				
$2\psi_1$	$\frac{1}{2}\psi_2$	$\frac{5}{3}\psi_3$	$\frac{5}{12}\psi_4$	$\frac{1}{10}\psi_5$	ψ_1	ψ_2	$\frac{5}{6}\psi_3$	$\frac{1}{12}\psi_4$	$\frac{1}{40}\psi_5$	$2\psi_1$	$3\psi_2$	$\frac{2}{3}\psi_3$	$\frac{7}{24}\psi_4$	$\frac{3}{20}\psi_5$
−9	+6	−42	+18	−6	−5	+15	−30	+6	−3	−11	+55	−33	+33	−33
−7	+2	+14	−22	+14	−4	+6	+6	−6	+6	−9	+25	+3	−27	+57
−5	−1	+35	−17	−1	−3	−1	+22	−6	+1	−7	+1	+21	−33	+21
−3	−3	+31	+3	−11	−2	−6	+23	−1	−4	−5	−17	+25	−13	−29
−1	−4	+12	+18	−6	−1	−9	+14	+4	−4	−3	−29	+19	+12	−44
+1	−4	−12	+18	+6	0	−10	0	+6	0	−1	−35	+7	+28	−20
+3	−3	−31	+3	+11	+1	−9	−14	+4	+4	+1	−35	−7	+28	+20
+5	−1	−35	−17	+1	+2	−6	−23	−1	+4	+3	−29	−19	+12	+44
+7	+2	−14	−22	−14	+3	−1	−22	−6	−1	+5	−17	−25	−13	+29
+9	+6	+42	+18	+6	+4	+6	−6	−6	−6	+7	+1	−21	−33	−21
					+5	+15	+30	+6	+3	+9	+25	−3	−27	−57
										+11	+55	+33	+33	+33
330	132	8580	2860	780	110	858	4290	286	156	572	12 012	5148	8008	15 912

TABLE II (continued)

13

ψ_1	ψ_2	$\frac{1}{6}\psi_3$	$\frac{7}{12}\psi_4$	$\frac{7}{120}\psi_5$
0	−14	0	+84	0
+1	−13	−4	+64	+20
+2	−10	−7	+11	−26
+3	−5	−8	−54	−11
+4	+2	−6	−96	−18
+5	+11	0	−66	−33
+6	+22	+11	+99	+22
182	2002	572	68 068	6188

14

$2\psi_1$	$\frac{1}{2}\psi_2$	$\frac{5}{3}\psi_3$	$\frac{7}{12}\psi_4$	$\frac{7}{30}\psi_5$
+1	−8	−24	+108	+60
+3	−7	−67	+63	+145
+5	−5	−95	−13	+139
+7	−2	−98	−92	+28
+9	+2	−66	−132	−132
+11	+7	+11	−77	−187
+13	+13	+143	+143	+143
910	728	97 240	136 136	235 144

15

ψ_1	$3\psi_2$	$\frac{5}{6}\psi_3$	$\frac{35}{12}\psi_4$	$\frac{21}{20}\psi_5$
0	−56	0	+756	0
+1	−53	−27	+621	+675
+2	−44	−49	+251	+1000
+3	−29	−61	−249	+751
+4	−8	−58	−704	−44
+5	+19	−35	−869	−979
+6	+52	+13	−429	−1144
+7	+91	+91	+1001	+1001
280	37 128	39 780	6 466 460	10 581 480

16

$2\psi_1$	ψ_2	$\frac{10}{3}\psi_3$	$\frac{7}{12}\psi_4$	$\frac{1}{10}\psi_5$
+1	−21	−63	+189	+45
+3	−19	−179	+129	+115
+5	−15	−265	+23	+131
+7	−9	−301	−101	+77
+9	−1	−267	−201	−33
+11	+9	−143	−221	−143
+13	+21	+91	−91	−143
+15	+35	+455	+273	+143
1360	5712	1 007 760	470 288	201 552

17

ψ_1	ψ_2	$\frac{1}{6}\psi_3$	$\frac{1}{12}\psi_4$	$\frac{1}{20}\psi_5$
0	−24	0	+36	0
+1	−23	−7	+31	+55
+2	−20	−13	+17	+88
+3	−15	−17	−3	+83
+4	−8	−18	−24	+36
+5	+1	−15	−39	−39
+6	+12	−7	−39	−104
+7	+25	+7	−13	−91
+8	+40	+23	+52	+104
408	7752	3876	16 796	100 776

18

$2\psi_1$	$\frac{3}{2}\psi_2$	$\frac{1}{3}\psi_3$	$\frac{1}{12}\psi_4$	$\frac{3}{10}\psi_5$
+1	−40	−8	+44	+220
+3	−37	−23	+33	+583
+5	−31	−35	+13	−733
+7	−22	−42	−12	+588
+9	−10	−42	−36	+156
+11	+5	−33	−51	−429
+13	+23	−13	−47	−871
+15	+44	+20	−12	−676
+17	+68	+68	+68	+884
1938	23 256	23 256	28 424	6 953 544

TABLE II (continued)

		19		
ψ_1	ψ_2	$\frac{5}{6}\psi_3$	$\frac{7}{12}\psi_4$	$\frac{1}{40}\psi_5$
0	−30	0	+396	0
+1	−29	−44	+352	+44
+2	−26	−83	+227	+74
+3	−21	−112	+42	+79
+4	−14	−126	−168	+54
+5	−5	−120	−354	+3
+6	+6	−89	−453	−58
+7	+19	−28	−388	−98
+8	+34	+68	−68	−68
+9	+51	+204	+612	+102
570	13 566	213 180	2 288 132	89 148

		20		
$2\psi_1$	ψ_2	$\frac{10}{3}\psi_3$	$\frac{35}{24}\psi_4$	$\frac{7}{20}\psi_5$
+1	−33	−99	+1188	+396
+3	−31	−287	+948	+1076
+5	−27	−445	+503	+1441
+7	−21	−553	−77	+1351
+9	−13	−591	−687	+771
+11	−3	−539	−1187	−187
+13	+9	−377	−1402	−1222
+15	+23	−85	−1122	−1802
+17	+39	+357	−102	−1122
+19	+57	+969	+1938	+1938
2660	17 556	4 903 140	22 881 320	31 201 800

		21		
ψ_1	$3\psi_2$	$\frac{5}{6}\psi_3$	$\frac{7}{12}\psi_4$	$\frac{21}{40}\psi_5$
0	−110	0	+594	0
+1	−107	−54	+540	+1404
+2	−98	−103	+385	+2444
+3	−83	−142	+150	+2819
+4	−62	−166	−130	+2354
+5	−35	−170	−406	+1063
+6	−2	−149	−615	−788
+7	+37	−98	−680	−2618
+8	+82	−12	−510	−3468
+9	+133	+114	0	−1938
+10	+190	+285	+969	+3876
770	201 894	432 630	5 720 330	121 687 020

		22		
$2\psi_1$	$\frac{1}{2}\psi_2$	$\frac{1}{3}\psi_3$	$\frac{7}{12}\psi_4$	$\frac{7}{30}\psi_5$
+1	−20	−12	+702	+390
+3	−19	−35	+585	+1079
+5	−17	−55	+365	+1509
+7	−14	−70	+70	+1554
+9	−10	−78	−258	+1158
+11	−5	−77	−563	+363
+13	+1	−65	−775	−663
+15	+8	−40	−810	−1598
+17	+16	0	−570	−1938
+19	+25	+57	+57	−969
+21	+35	+133	+1197	+2261
3542	7084	96 140	8 748 740	40 562 340

		23		
ψ_1	ψ_2	$\frac{1}{6}\psi_3$	$\frac{7}{12}\psi_4$	$\frac{1}{60}\psi_5$
0	−44	0	+858	0
+1	−43	−13	+793	+65
+2	−40	−25	+605	+116
+3	−35	−35	+315	+141
+4	−28	−42	−42	+132
+5	−19	−45	−417	+87
+6	−8	−43	−747	+12
+7	+5	−35	−955	−77
+8	+20	−20	−950	−152
+9	+37	+3	−627	−171
+10	+56	+35	+133	−76
+11	+77	+77	+1463	+209
1012	35 420	32 890	13 123 110	340 860

		24		
$2\psi_1$	$3\psi_2$	$\frac{10}{3}\psi_3$	$\frac{1}{12}\psi_4$	$\frac{3}{10}\psi_5$
+1	−143	−143	+143	+715
+3	−137	−419	+123	+2005
+5	−125	−665	+85	+2893
+7	−107	−861	+33	+3171
+9	−83	−987	−27	+2721
+11	−53	−1023	−87	+1551
+13	−17	−949	−137	−169
+15	+25	−745	−165	−2071
+17	+73	−391	−157	−3553
+19	+127	+133	−97	−3743
+21	+187	+847	+33	−1463
+23	+253	+1771	+253	+4807
4600	394 680	17 760 600	394 680	177 928 920

TABLE II (continued)

		25		
ψ_1	ψ_2	$\frac{5}{6}\psi_3$	$\frac{5}{12}\psi_4$	$\frac{1}{20}\psi_5$
0	−52	0	+858	0
+1	−51	−77	+803	+275
+2	−48	−149	+643	+500
+3	−43	−211	+393	+631
+4	−36	−258	+78	+636
+5	−27	−235	−267	+501
+6	−16	−287	−597	+236
+7	−3	−259	−857	−119
+8	+12	−196	−982	−483
+9	+29	−93	−897	−753
+10	+48	+55	−517	−748
+11	+69	+253	+253	−253
+12	+92	+506	+1518	+1012
1300	53 820	1 480 050	14 307 150	7 803 900

		26		
$2\psi_1$	$\frac{1}{2}\psi_2$	$\frac{5}{3}\psi_3$	$\frac{7}{12}\psi_4$	$\frac{1}{10}\psi_5$
+1	−28	−84	+1386	+330
+3	−27	−247	+1221	+935
+5	−25	−395	+905	+1381
+7	−22	−518	+466	+1582
+9	−18	−606	−54	+1482
+11	−13	−649	−599	+1067
+13	−7	−637	−1099	+377
+15	0	−560	−1470	−482
+17	+8	−408	−1614	−1326
+19	+17	−171	−1419	−1881
+21	+27	+161	−759	−1771
+23	+38	−598	+506	−506
+25	+50	+1150	+2530	+2530
5850	16 380	7 803 900	40 060 020	48 384 180

		27		
ψ_1	$3\psi_2$	$\frac{1}{6}\psi_3$	$\frac{7}{12}\psi_4$	$\frac{21}{40}\psi_5$
0	−182	0	+1638	0
+1	−179	−18	+1548	+3 960
+2	−170	−35	+1285	+7 304
+3	−155	−50	+870	+9 479
+4	−134	−62	+338	+10 058
+5	−107	−70	−262	+8 803
+6	−74	−73	−867	+5 728
+7	−35	−70	−1400	+1 162
+8	+10	−60	−1770	−4 188
+9	+61	−42	−1872	−9 174
+10	+118	−15	−1587	−12 144
+11	+181	+22	−782	−10 879
+12	+250	+70	+690	−2 530
+13	+325	+130	+2990	+16 445
1638	712 530	101 790	56 448 210	2 032 135 560

		28		
$2\psi_1$	ψ_2	$\frac{2}{3}\psi_3$	$\frac{7}{24}\psi_4$	$\frac{7}{20}\psi_5$
+1	−65	−39	+936	+1 560
+3	−63	−115	+840	+4 456
+5	−59	−185	+655	+6 701
+7	−53	−245	+395	+7 931
+9	−45	−291	+81	+7 887
+11	−35	−319	−259	+6 457
+13	−23	−325	−590	+3 718
+15	−9	−305	−870	−22
+17	+7	−255	−1050	−4 182
+19	+25	−171	−1074	−7 866
+21	+45	−49	−879	−9 821
+23	+67	+115	−395	−8 395
+25	+91	+325	+455	−1 495
+27	+117	+585	+1755	+13 455
7309	95 004	2 103 660	19 634 160	1 354 757 040

TABLE II (continued)

29				
ψ_1	ψ_2	$\frac{5}{6}\psi_3$	$\frac{7}{12}\psi_4$	$\frac{7}{40}\psi_5$
0	− 70	0	+2184	0
+ 1	− 69	−104	+2080	+1768
+ 2	− 66	−203	+1775	+3298
+ 3	− 61	−292	+1290	+4373
+ 4	− 54	−366	+660	+4818
+ 5	− 45	−420	−66	+4521
+ 6	− 34	−449	−825	+3454
+ 7	− 21	−448	−1540	+1694
+ 8	− 6	−412	−2120	−556
+ 9	+ 11	−336	−2460	−2946
+10	+ 30	−215	−2441	−4958
+11	+ 51	− 44	−1930	−5885
+12	+.74	+182	−780	−4810
+13	+ 99	+468	+1170	−585
+14	+126	+819	+4095	+8190
2030	113 274	4 207 320	107 987 880	500 671 080

30				
$2\psi_1$	$\frac{3}{2}\psi_2$	$\frac{5}{3}\psi_3$	$\frac{35}{12}\psi_4$	$\frac{3}{10}\psi_5$
+1	−112	−112	+12376	+1768
+3	−109	−331	+11271	+5083
+5	−103	−535	+9131	+7753
+7	−94	−714	+6096	+9408
+9	−82	−858	+2376	+9768
+11	−67	−957	−1749	+8679
+13	−49	−1001	−5929	+6149
+15	−28	−980	−9744	+2384
+17	−4	−884	−12704	−2176
+19	+23	−703	−14249	−6821
+21	+53	−427	−13749	−10535
+23	+86	−46	−10504	−11960
+25	+122	+450	−3744	−9360
+27	+161	+1071	+7341	−585
+29	+203	+1827	+23751	+16965
8990	302 064	21 360 240	3 671 587 920	2 145 733 200

31				
ψ_1	ψ_2	$\frac{5}{6}\psi_3$	$\frac{1}{12}\psi_4$	$\frac{1}{60}\psi_5$
0	−80	0	+408	0
+1	−79	−119	+391	+221
+2	−76	−233	+341	+416
+3	−71	−337	+261	+561
+4	−64	−426	+156	+636
+5	−55	−495	+33	+627
+6	−44	−539	−99	+528
+7	−31	−553	−229	+343
+8	−16	−532	−344	+88
+9	+1	−471	−429	−207
+10	+20	−365	−467	−496
+11	+41	−209	−439	−715
+12	+64	+2	−324	−780
+13	+89	+273	−99	−585
+14	+116	+609	+261	0
+15	+145	+1015	+783	+1131
2480	158 224	6 724 520	4 034 712	9 536 592

32				
$2\psi_1$	ψ_2	$\frac{2}{3}\psi_3$	$\frac{1}{12}\psi_4$	$\frac{1}{30}\psi_5$
+1	−85	−51	+459	+255
+3	−83	−151	+423	+737
+5	−79	−245	+353	+1137
+7	−73	−329	+253	+1407
+9	−65	−399	+129	+1509
+11	−55	−451	−11	+1419
+13	−43	−481	−157	+1131
+15	−29	−485	−297	+661
+17	−13	−459	−417	+51
+19	+5	−399	−501	−627
+21	+25	−301	−531	−1267
+23	+47	−161	−487	−1725
+25	+71	+25	−347	−1815
+27	+97	+261	−87	−1305
+29	+125	+551	+319	−87
+31	+155	+899	+899	+2697
10 912	185 504	5 379 616	5 379 616	54 285 216

TABLE II (continued)

33

ψ_1	$3\psi_2$	$\frac{1}{6}\,\psi_3$	$\frac{7}{12}\,\psi_4$	$\frac{3}{20}\,\psi_5$
0	−272	0	+3672	0
+1	−269	−27	+3537	+2565
+2	−260	−53	+3139	+4864
+3	−245	−77	+2499	+6649
+4	−224	−98	+1652	+7708
+5	−197	−115	+647	+7883
+6	−164	−127	−453	+7088
+7	−125	−133	−1571	+5327
+8	−80	−132	−2616	+2712
+9	−29	−123	−3483	−519
+10	+28	−105	−4053	−3984
+11	+91	−77	−4193	−7139
+12	+160	−38	−3756	−9260
+13	+235	+13	−2581	−9425
+14	+316	+77	−493	−6496
+15	+403	+155	+2697	+899
+16	+496	+248	+7192	+14384
2992	1 947 792	417 384	348 330 136	1 547 128 656

34

$2\psi_1$	$\frac{1}{2}\,\psi_2$	$\frac{5}{3}\,\psi_3$	$\frac{7}{12}\,\psi_4$	$\frac{7}{10}\,\psi_5$
+1	−48	−144	+4104	+6840
+3	−47	−427	+3819	+19855
+5	−45	−695	+3263	+30917
+7	−42	−938	+2464	+38864
+9	−38	−1146	+1464	+42744
+11	−33	−1309	+319	+41899
+13	−27	−1417	−901	+36049
+15	−20	−1460	−2112	+23376
+17	−12	−1428	−3216	+10608
+19	−3	−1311	−4101	−6897
+21	+7	−1099	−4641	−25067
+23	+18	−782	−4696	−41032
+25	+30	−350	−4112	−51040
+27	+43	+207	−2721	−50373
+29	+57	+899	−341	−33263
+31	+72	+1736	+3224	+7192
+33	+88	+2728	+8184	+79112
13090	62 832	51 477 360	456 432 592	46 929 569 232

35

ψ_1	ψ_2	$\frac{5}{6}\,\psi_3$	$\frac{35}{12}\,\psi_4$	$\frac{7}{40}\,\psi_5$
0	−102	0	+23256	0
+1	−101	−152	+22496	+3800
+2	−98	−299	+20251	+7250
+3	−93	−436	+16626	+10021
+4	−86	−558	+11796	+11826
+5	−77	−660	+6006	+12441
+6	−66	−737	−429	+11726
+7	−53	−784	−7124	+9646
+8	−38	−796	−13624	+6292
+9	−21	−768	−19404	+1902
+10	−2	−695	−23869	−3118
+11	+19	−572	−26354	−8173
+12	+42	−394	−26124	−12458
+13	+67	−156	−22374	−14937
+14	+94	+147	−14229	−14322
+15	+123	+520	−744	−9052
+16	+154	+968	+19096	+2723
+17	+187	+1496	+46376	+23188
3570	290 598	15 775 320	14 834 059 240	4 045 652 520

36

$2\psi_1$	$3\psi_2$	$\frac{10}{3}\,\psi_3$	$\frac{7}{24}\,\psi_4$	$\frac{21}{20}\,\psi_5$
+1	−323	−323	+2584	+12 920
+3	−317	−959	+2424	+37 640
+5	−305	−1565	+2111	+59 063
+7	−287	−2121	+1659	+75 201
+9	−263	−2607	+1089	+84 381
+11	−233	−3003	+429	+85 371
+13	−197	−3259	−236	+77 506
+15	−155	−3445	−1014	+60 814
+17	−107	−3451	−1706	+36 142
+19	−53	−3287	−2306	+5 282
+21	+7	−2933	−2751	−28 903
+23	+73	−2369	−2971	−62 353
+25	+145	−1575	−2889	−89 685
+27	+223	−531	−2421	−104 067
+29	+307	+783	−1476	−97 092
+31	+397	+2387	+44	−58 652
+33	+493	+4301	+2244	−23 188
+35	+595	+6545	+5236	+162 316
15540	3 011 652	307 618 740	191 407 216	199 046 103 984

TABLE II (continued)

		37		
ψ_1	ψ_2	$\frac{1}{6}\psi_3$	$\frac{7}{12}\psi_4$	$\frac{1}{40}\psi_5$
0	−114	0	+5 814	0
+1	−113	−34	+5 644	+680
+2	−110	−67	+5 141	+1 304
+3	−105	−98	+4 326	+1 819
+4	−98	−126	+3 234	+2 178
+5	−89	−150	+1 914	+2 343
+6	−78	−169	+429	+2 288
+7	−65	−182	−1 144	+2 002
+8	−50	−188	−2 714	+1 492
+9	−33	−186	−4 176	+786
+10	−14	−175	−5 411	−64
+11	+7	−154	−6 286	−979
+12	+30	−122	−6 654	−1 850
+13	+55	−78	−6 354	−2 535
+14	+82	−21	−5 211	−2 856
+15	+111	+50	−3 036	−2 596
+16	+142	+136	+374	−1 496
+17	+175	+238	+5 236	−748
+18	+210	+357	+11 781	+4 488
4218	383 838	932 178	980 961 982	152 877 192

		38		
$2\psi_1$	$\frac{1}{2}\varphi_2$	$\frac{1}{3}\psi_3$	$\frac{1}{12}\psi_4$	$\frac{1}{10}\psi_5$
+1	−60	−36	+918	+1 530
+3	−59	−107	+867	+4 471
+5	−57	−175	+767	+7 061
+7	−54	−238	+622	+9 086
+9	−50	−294	+438	+10 362
+11	−45	−341	+223	+10 747
+13	−39	−377	−13	+10 153
+15	−32	−400	−258	+8 558
+17	−21	−408	−498	+6 018
+19	−15	−399	−717	+2 679
+21	−5	−371	−897	−1 211
+23	+6	−322	−1018	−5 290
+25	+18	−250	−1058	−9 070
+27	+31	−153	−993	−11 925
+29	+45	−29	−797	−13 079
+31	+60	+124	−442	−11 594
+33	+76	+308	+102	−6 358
+35	+93	+525	+867	+3 927
+37	+111	+777	+1887	+20 757
18 278	109 668	4 496 388	25 479 532	3 286 859 628

		39		
ψ_1	$3\psi_2$	$\frac{5}{6}\psi_3$	$\frac{1}{12}\psi_4$	$\frac{3}{20}\psi_5$
0	−380	0	+1026	0
+1	−377	−189	+999	+5 049
+2	−368	−373	+919	+9 724
+3	−353	−547	+789	+13 669
+4	−332	−706	+614	+16 564
+5	−305	−845	+401	+18 143
+6	−272	−959	+159	+18 212
+7	−233	−1043	−101	+16 667
+8	−188	−1092	−366	+13 512
+9	−137	−1101	−621	+8 877
+10	−80	−1065	−849	+3 036
+11	−17	−979	−1031	−3 575
+12	+52	−833	−1146	−10 340
+13	+127	−637	−1171	−16 445
+14	+208	−371	−1081	−20 860
+15	+295	−35	−849	−22 321
+16	+388	+376	−446	−19 312
+17	+487	+867	+159	−10 047
+18	+592	+1443	+999	+7 548
+19	+703	+2109	+2109	+35 853
4940	4 496 388	33 722 910	32 224 114	9 860 578 884

		40		
$2\psi_1$	ψ_2	$\frac{10}{3}\psi_3$	$\frac{35}{12}\psi_4$	$\frac{1}{30}\psi_5$
+1	−133	−399	+39 501	+627
+3	−131	−1187	+37 521	+1837
+5	−127	−1945	+33 631	+2917
+7	−121	−2653	+27 971	+3787
+9	−113	−3291	+20 751	+4377
+11	−103	−3839	+12 251	+4631
+13	−91	−4277	+2 821	+4511
+15	−77	−4585	−7 119	+4001
+17	−61	−4743	−17 079	+3111
+19	−43	−4731	−26 499	+1881
+21	−23	−4529	−34 749	+385
+23	−1	−4117	−41 129	−1265
+25	+23	−3475	−44 869	−2915
+27	+49	−2583	−45 129	−4365
+29	+77	−1421	−40 999	−5365
+31	+107	+31	−31 499	−5611
+33	+139	+1793	−15 579	−4741
+35	+173	+3885	−7 881	−2331
+37	+209	+6327	+40 071	+2109
+39	+247	+9139	+82 251	+9139
21 320	567 112	644 482 280	49 625 135 560	644 482 280

TABLE II (continued)

41

ψ_1	ψ_2	$\frac{5}{6}\,\psi_3$	$\frac{7}{12}\,\psi_4$	$\frac{7}{60}\,\psi_5$
0	−140	0	+8 778	0
+1	−139	−209	+8 569	+4 807
+2	−136	−413	+7 949	+9 292
+3	−131	−607	+6 939	+13 147
+4	−124	−786	+5 574	+16 092
+5	−115	−945	+3 903	+17 889
+6	−104	−1079	+1 989	+18 356
+7	−91	−1183	−91	+17 381
+8	−76	−1252	−2 246	+14 936
+9	−59	−1281	−4 371	+11 091
+10	−40	−1265	−6 347	+6 028
+11	−19	−1199	−8 041	+55
+12	+4	−1078	−9 306	−6 380
+13	+29	−897	−9 981	−12 675
+14	+56	−651	−9 891	−18 060
+15	+85	−335	−8 847	−21 583
+16	+116	+56	−6 646	−22 096
+17	+149	+527	−3 071	−18 241
+18	+184	+1083	+2 109	−8 436
+19	+221	+1729	+9 139	+9 139
+20	+260	+2470	+18 278	+36 556
5740	641 732	47 900 710	2 481 256 778	10 376 164 708

42

$2\psi_1$	$\frac{3}{2}\,\psi_2$	$\frac{1}{3}\,\psi_3$	$\frac{7}{12}\,\psi_4$	$\frac{21}{10}\,\psi_5$
+1	−220	−44	+9 614	+48 070
+3	−217	−131	+9 177	+141 151
+5	−211	−215	+8 317	−225 181
+7	−202	−294	+7 062	−294 546
+9	−190	−366	+5 454	−344 262
+11	−175	−429	+3 549	−370 227
+13	−157	−481	+1 417	−369 473
+15	−136	−520	−858	−340 418
+17	−112	−544	−3 178	−283 118
+19	−85	−551	−5 431	+199 519
+21	−55	−539	−7 491	+93 709
+23	−22	−506	−9 218	−27 830
+25	+14	−450	−10 458	−155 970
+27	+53	−369	−11 043	−278 685
+29	+95	−261	−10 791	−380 799
+31	+140	−124	−9 506	−443 734
+33	+188	+44	−6 978	−445 258
+35	+239	+245	−2 983	−359 233
+37	+293	+481	+2 717	−155 363
+39	+350	+754	+10 374	+201 058
+41	+410	+1066	+20 254	+749 398
24 682	1 629 012	9 075 924	3 084 805 724	4 389 117 671 484

43

ψ_1	ψ_2	$\frac{1}{6}\,\psi_3$	$\frac{7}{12}\,\psi_4$	$\frac{7}{40}\,\psi_5$
0	−154	0	+10 626	0
+1	−153	−46	+10 396	+8 740
+2	−150	−91	+9 713	+16 948
+3	−149	−134	+8 598	+24 113
+4	−138	−174	+7 086	+29 766
+5	−129	−210	+5 226	+33 501
+6	−118	−241	+3 081	+34 996
+7	−105	−266	+728	+34 034
+8	−90	−284	−1 742	+30 524
+9	−73	−294	−4 224	+24 522
+10	−54	−295	−6 599	+16 252
+11	−33	−286	−8 734	+6 127
+12	−10	−266	−10 482	−5 230
+13	+15	−234	−11 682	−16 965
+14	+42	−189	−12 159	−27 972
+15	+71	−130	−11 724	−36 872
+16	+102	−56	−10 174	−41 992
+17	+135	+34	−7 292	−41 344
+18	+170	+141	−2 847	−32 604
+19	+207	+266	+3 406	−13 091
+20	+246	+410	+11 726	+20 254
+21	+287	+574	+22 386	+70 889
6622	814 506	2 676 234	3 815 417 606	39 541 600 644

44

$2\psi_1$	ψ_2	$\frac{10}{3}\,\psi_3$	$\frac{7}{24}\,\psi_4$	$\frac{1}{20}\,\psi_5$
+1	−161	−483	+5 796	+1 380
+3	−159	−1 439	+5 556	−4 060
+5	−155	−2 365	+5 083	−6 503
+7	−149	−3 241	+4 391	−8 561
+9	−141	−4 047	+3 501	−10 101
+11	−131	−4 763	+2 441	−11 011
+13	−119	−5 369	+1 246	−11 206
+15	−105	−5 845	−42	−10 634
+17	−89	−6 171	−1 374	−9 282
+19	−71	−6 327	−2 694	−7 182
+21	−51	−6 293	−3 939	−4 417
+23	−29	−6 049	−5 039	−1 127
+25	−5	−5 575	−5 917	−2 485
+27	+21	−4 851	−6 489	−6 147
+29	+49	−3 857	−6 664	−9 512
+31	+79	−2 573	−6 344	−12 152
+33	+111	−979	−5 424	−13 552
+35	+145	+945	−3 792	−13 104
+37	+181	+3 219	−1 329	−10 101
+39	+219	+5 863	+2 091	−3 731
+41	+259	+8 897	+6 601	+6 929
+43	+301	+12 341	+12 341	+22 919
28 380	913 836	1 257 829 980	1 173 974 648	4 162 273 752

TABLE II (continued)

45				
ψ_1	$3\psi_2$	$\dfrac{5}{6}\psi_3$	$\dfrac{5}{12}\psi_4$	$\dfrac{3}{40}\psi_5$
0	−506	0	+9 108	0
+1	−503	−252	+8 928	+4 500
+2	−494	−499	+8 393	+8 750
+3	−479	−736	+7 518	+12 509
+4	−458	−958	+6 328	+15 554
+5	−431	−1160	+4 858	+17 689
+6	−398	−1337	+3 153	+18 754
+7	−359	−1484	+1 268	+18 634
+8	−314	−1596	−732	+17 268
+9	−263	−1668	−2 772	+14 658
+10	−206	−1695	−4 767	+10 878
+11	−143	−1672	−6 622	+6 083
+12	−74	−1594	−8 232	+518
+13	+1	−1456	−9 482	−5 473
+14	+82	−1253	−10 247	−11 438
+15	+169	−980	−10 392	−16 808
+16	+262	−632	−9 772	−20 888
+17	+361	−204	−8 232	−22 848
+18	+466	+309	−5 607	−21 714
+19	+577	+912	−1 722	−16 359
+20	+694	+1610	+3 608	−5 494
+21	+817	+2408	+10 578	+12 341
+22	+946	+3311	+19 393	+38 786
7590	9 203 634	92 036 340	2 934 936 620	12 006 558 900

46				
$2\psi_1$	$\dfrac{1}{2}\psi_2$	$\dfrac{5}{3}\psi_3$	$\dfrac{1}{12}\psi_4$	$\dfrac{1}{10}\psi_5$
+1	−88	−264	+1980	+3 300
+3	−87	−787	+1905	+9 725
+5	−85	−1295	+1757	+15 631
+7	−82	−1778	+1540	+20 692
+9	−78	−2226	+1260	+24 612
+11	−73	−2629	+925	+27 137
+13	−67	−2977	+545	+28 067
+15	−60	−3260	+132	+27 268
+17	−52	−3468	−300	+24 684
+19	−43	−3591	−735	+20 349
+21	−33	−3619	−1155	+14 399
+23	−22	−3542	−1540	+7 084
+25	−10	−3350	−1868	−1 220
+27	+3	−3033	−2115	−9 999
+29	+17	−2581	−2255	−18 589
+31	+32	−1984	−2260	−26 164
+33	+48	−1232	−2100	−31 724
+35	+65	−315	−1743	−34 083
+37	+83	+777	−1155	−31 857
+39	+102	+2054	−300	−23 452
+41	+122	+3526	+860	−7 052
+43	+143	+5203	+2365	+19 393
+45	+165	+7095	+4257	+58 179
32 430	285 384	429 502 920	143 167 640	27 214 866 840

TABLE II (continued)

47				
ψ_1	ψ_2	$\frac{1}{6}\psi_3$	$\frac{7}{12}\psi_4$	$\frac{1}{20}\psi_5$
0	−184	0	+15 180	0
+1	−183	−55	+14 905	+3 575
+2	−180	−109	+14 087	+6 968
+3	−175	−161	+12 747	+10 003
+4	−168	−210	+10 920	+12 516
+5	−159	−255	+8 655	+14 361
+6	−148	−295	+6 015	+15 416
+7	−135	−329	+3 077	+15 589
+8	−120	−356	−68	+14 824
+9	−103	−375	−3 315	+13 107
+10	−84	−385	−6 545	+10 472
+11	−63	−385	−9 625	+7 007
+12	−40	−374	−12 408	+2 860
+13	−15	−351	−14 733	−1 755
+14	+12	−315	−16 425	−6 552
+15	+41	−265	−17 295	−11 167
+16	+72	−200	−17 140	−15 152
+17	+105	−119	−15 743	−17 969
+18	+140	−21	−12 873	−18 984
+19	+177	+95	−8 285	−17 461
+20	+216	+230	−1 720	−12 556
+21	+257	+385	+7 095	−3 311
+22	+300	+561	+18 447	+11 352
+23	+345	+759	+32 637	+32 637
8648	1 271 256	4 994 220	8 518 474 580	8 629 104 120

48				
$2\psi_1$	$3\psi_2$	$\frac{2}{3}\psi_3$	$\frac{7}{12}\psi_4$	$\frac{21}{10}\psi_5$
+1	−575	−115	+16 445	+82 225
+3	−569	−343	+15 873	+242 671
+5	−557	−565	+14 743	+391 231
+7	−539	−777	+13 083	+520 401
+9	−515	−975	+10 935	+623 307
+11	−485	−1155	+8 355	+693 957
+13	−449	−1313	+5 413	+727 493
+15	−407	−1445	+2 193	+720 443
+17	−359	−1547	−1 207	+670 973
+19	−305	−1615	−4 675	+579 139
+21	−245	−1645	−8 085	+447 139
+23	−179	−1633	−11 297	+279 565
+25	−107	−1575	−14 157	+83 655
+27	−29	−1467	−16 497	−130 455
+29	+55	−1305	−18 135	−349 479
+31	+145	−1085	−18 875	−556 729
+33	+241	−803	−18 507	−731 863
+35	+343	−455	−16 807	−850 633
+37	+451	−37	−13 537	−884 633
+39	+565	+455	−8 445	−801 047
+41	+685	+1025	−1 265	−562 397
+43	+811	+1677	+8 283	−126 291
+45	+943	+2415	+20 493	+554 829
+47	+1081	+3243	+35 673	+1 533 939
36 848	12 712 560	92 620 080	10 301 411 120	19 208 385 771 120

TABLE II (continued)

49				
ψ_1	ψ_3	$\dfrac{5}{6}\,\psi_3$	$\dfrac{7}{12}\,\psi_4$	$\dfrac{7}{60}\,\psi_5$
0	−200	0	+17 940	0
+1	−199	−299	+17 641	+9 867
+2	−196	−593	+16 751	+19 272
+3	−191	−877	+15 291	+27 767
+4	−184	−1146	+13 296	+34 932
+5	−175	−1395	+10 815	+40 389
+6	−164	−1619	+7 911	+43 816
+7	−151	−1813	+4 661	+44 961
+8	−136	−1972	+1 156	+43 656
+9	−119	−2091	−2 499	+39 831
+10	−100	−2165	−6 185	+33 528
+11	−79	−2189	−9 769	+24 915
+12	−56	−2158	−13 104	+14 300
+13	−31	−2067	−16 029	+2 145
+14	−4	−1911	−18 369	−10 920
+15	+25	−1685	−19 935	−24 083
+16	+56	−1384	−20 524	−36 336
+17	+89	−1003	−19 919	−46 461
+18	+124	−537	−17 889	−53 016
+19	+161	+19	−14 189	−54 321
+20	+200	+670	−8 560	−48 444
+21	+241	+1421	−729	−33 187
+22	+284	+2277	+9 591	−6 072
+23	+329	+3243	+22 701	+35 673
+24	+376	+4324	+38 916	+95 128
9800	1 566 040	167 230 700	12 408 517 940	74 451 107 640

50				
$2\psi_1$	$\dfrac{1}{2}\,\psi_2$	$\dfrac{5}{3}\,\psi_3$	$\dfrac{35}{12}\,\psi_4$	$\dfrac{7}{30}\,\psi_5$
+1	−104	−312	+96 876	+10 764
+3	−103	−931	+93 771	+31 809
+5	−101	−1535	+87 631	+51 419
+7	−98	−2114	+78 596	+68 684
+9	−94	−2658	+66 876	+82 764
+11	−89	−3157	+52 751	+92 917
+13	−83	−3601	+36 571	+98 527
+15	−76	−3980	+18 756	+99 132
+17	−68	−4284	−204	+94 452
+19	−59	−4503	−19 749	+84 417
+21	−49	−4627	−39 249	+69 195
+23	−38	−4646	−58 004	+49 220
+25	−26	−4550	−75 244	+25 220
+27	−13	−4329	−90 129	−1 755
+29	+1	−3973	−101 749	−30 305
+31	+16	−3472	−109 124	−58 652
+33	+32	−2816	−111 204	−84 612
+35	+49	−1995	−106 869	−105 567
+37	+67	−999	−94 929	−118 437
+39	+86	+182	−74 124	−219 652
+41	+106	+1558	−43 124	−105 124
+43	+127	+3139	−529	−70 219
+45	+149	+4935	+55 131	−9 729
+47	+172	+6956	+125 396	+82 156
+49	+196	+9212	+211 876	+211 876
41650	433 160	770 715 400	372 255 538 200	372 255 538 200

TABLE III. Sums of Powers of Deviations from the Mean Value and Determinants of Distribution of Series of Integers

n	Σx^2	Σx^4	Σx^6	Σx^8	D_2	D_3	D_4
1	0	0	0	0	0	0	0
2	0,5	0,125	0,03125	0,0078125	0	0	0
3	2	2	2	2	2	0	0
4	5	10,250	22,81250	51,2656250	16	9	0
5	10	34	130	514	70	144	576
6	17,5	88,375	511,09375	3 103,0234375	224	1 134	18 432
7	28	196	1 588	13 636	588	6 048	266 112
8	42	388,500	4 187,62500	48 140,5312500	1 344	24 948	2 433 024
9	60	708	9 780	144 708	2 772	85 536	16 308 864
10	82,5	1208,625	20 795,15625	384 443,0390625	5 280	254 826	86 980 608
11	110	1958	41 030	925 958	9 438	679 536	388 694 592
12	143	3038,750	76 156,43750	2 059 121,7968750	16 016	1 656 369	1 507 663 872
13	182	4550	134 342	4 285 190	26 026	3 747 744	5 206 151 808
14	227,5	6608,875	226 994,21875	8 432 018,0546875	40 768	7 963 956	16 310 181 888
15	280	9352	369 640	15 814 792	61 880	16 039 296	47 037 399 552
16	340	12 937	582 951,25000	28 454 601,0625000	91 392	30 837 456	126 310 219 776
17	408	17 544	893 928	49 369 224	131 784	56 930 688	318 735 945 216
18	484,5	23 377,125	1 337 250,28125	82 952 706,0703125	186 048	101 407 788	761 504 882 688
19	570	30 666	1 956 810	135 462 666	257 754	174 978 144	1 733 216 842 368
20	665	39 667,250	2 807 434,06250	215 636 792,3281250	351 120	293 452 929	3 777 661 476 864
21	770	50 666	3 956 810	335 462 666	471 086	479 700 144	7 919 316 377 280
22	885,5	63 977,375	5 487 625,34375	511 127 881,0859375	623 392	766 187 730	16 027 771 668 480
23	1012	79 948	7 499 932	764 180 428	814 660	1 198 248 480	31 418 075 145 600
24	1150	98 957,500	10 113 746,37500	1 122 932 453,5937500	1 052 480	1 838 222 100	59 816 564 121 600
25	1300	121 420	13 471 900	1 624 143 820	1 345 500	2 770 653 600	110 881 557 072 000

TABLE IV. Sums of Powers of Integers

n	$S(n)$	$S(n^2)$	$S(n^3)$	$S(n^4)$	$S(n^5)$	$S(n^6)$	$S(n^7)$	n
1	1	1	1	1	1	1	1	1
2	3	5	9	17	33	65	129	2
3	6	14	36	98	276	794	2 316	3
4	10	30	100	354	1300	4 890	18 700	4
5	15	55	225	979	4425	20 515	96 825	5
6	21	91	441	2 275	12 201	67 171	376 761	6
7	28	140	784	4 676	29 008	184 820	1 200 304	7
8	36	204	1 296	8 772	61 776	446 964	3 297 456	8
9	45	285	2 025	15 333	120 825	978 405	8 080 425	9
10	55	385	3 025	25 333	220 825	1 978 405	18 080 425	10
11	66	506	4 356	39 974	381 876	3 749 966	37 567 596	11
12	78	650	6 084	60 710	630 708	6 735 950	73 399 404	12
13	91	819	8 281	89 271	1 002 001	11 562 759	136 147 921	13
14	105	1 015	11 025	127 687	1 539 825	19 092 295	241 561 425	14
15	120	1 240	14 400	178 312	2 299 200	30 482 920	412 420 800	15
16	136	1 496	18 496	243 848	3 347 776	47 260 136	680 856 256	16
17	153	1 785	23 409	327 369	4 767 633	71 397 705	1 091 194 929	17
18	171	2 109	29 241	432 345	6 657 201	105 409 929	1 703 414 961	18
19	190	2 470	36 100	562 666	9 133 300	152 455 810	2 597 286 700	19
20	210	2 870	44 100	722 666	12 333 300	216 455 810	3 877 286 700	20
21	231	3 311	53 361	917 147	16 417 401	302 221 931	5 678 375 241	21
22	253	3 795	64 009	1 151 403	21 571 033	415 601 835	8 172 733 129	22
23	276	4 324	76 176	1 431 244	28 007 376	563 637 724	11 577 558 576	23
24	300	4 900	90 000	1 763 020	35 970 000	754 740 700	16 164 030 000	24
25	325	5 525	105 625	2 153 645	45 735 625	998 881 325	22 267 545 625	25
26	351	6 201	123 201	2 610 621	57 617 001	1 307 797 101	30 299 355 801	26
27	378	6 930	142 884	3 142 062	71 965 908	1 695 217 590	40 759 709 004	27
28	406	7 714	164 836	3 756 718	89 176 276	2 177 107 894	54 252 637 516	28
29	435	8 555	189 225	4 463 999	109 687 425	2 771 931 215	71 502 513 825	29
30	465	9 455	216 225	5 273 999	133 987 425	3 500 931 215	93 372 513 825	30

TABLE IV (continued)

n	$S(n)$	$S(n^2)$	$S(n^3)$	$S(n^4)$	$S(n^5)$	$S(n^6)$	$S(n^7)$	n
31	496	10 416	246 016	6 197 520	162 616 576	4 388 434 896	120 885 127 936	31
32	528	11 440	278 784	7 246 096	196 171 008	5 462 176 720	155 244 866 304	32
33	561	12 529	314 721	8 432 017	235 306 401	6 753 644 689	197 863 309 281	33
34	595	13 685	354 025	9 768 353	280 741 825	8 298 449 105	250 386 659 425	34
35	630	14 910	396 900	11 268 978	333 263 700	10 136 714 730	314 725 956 300	35
36	666	16 206	443 556	12 948 594	393 729 876	12 313 497 066	393 090 120 396	36
37	703	17 575	494 209	14 822 755	463 073 833	14 879 223 475	488 021 997 529	37
38	741	19 019	549 081	16 907 891	542 309 001	17 890 159 859	602 437 580 121	38
39	780	20 540	608 400	19 221 332	632 533 200	21 408 903 620	739 668 586 800	39
40	820	22 140	672 400	21 781 332	734 933 200	25 504 903 620	903 508 586 800	40
41	861	23 821	741 321	24 607 093	850 789 401	30 255 007 861	1 098 262 860 681	41
42	903	25 585	815 409	27 718 789	981 480 633	35 744 039 605	1 328 802 193 929	42
43	946	27 434	894 916	31 137 590	1 128 489 076	42 065 402 654	1 600 620 805 036	43
44	990	29 370	980 100	34 885 686	1 293 405 300	49 321 716 510	1 919 898 614 700	44
45	1035	31 395	1 071 225	38 986 311	1 477 933 425	57 625 482 135	2 293 568 067 825	45
46	1081	33 511	1 168 561	43 463 767	1 683 896 401	67 099 779 031	2 729 385 725 041	46
47	1128	35 720	1 272 384	48 343 448	1 913 241 408	77 878 994 360	3 236 008 845 504	47
48	1176	38 024	1 382 976	53 651 864	2 168 045 376	90 109 584 824	3 823 077 187 776	48
49	1225	40 425	1 500 625	59 416 665	2 450 520 625	103 950 872 025	4 501 300 260 625	49
50	1275	42 925	1 625 625	65 666 665	2 763 020 625	119 575 872 025	5 282 550 260 625	50
51	1326	45 526	1 758 276	72 431 866	3 108 045 876	137 172 159 826	6 179 960 938 476	51
52	1378	48 230	1 898 884	79 743 482	3 488 249 908	156 942 769 490	7 208 032 641 004	52
53	1431	51 039	2 047 761	87 633 963	3 906 445 401	179 107 130 619	8 382 743 780 841	53
54	1485	53 955	2 205 225	96 137 019	4 365 610 425	203 902 041 915	9 721 668 990 825	54
55	1540	56 980	2 371 600	105 287 644	4 868 894 800	231 582 682 540	11 244 104 225 200	55
56	1596	60 116	2 547 216	115 122 140	5 419 626 576	262 423 661 996	12 971 199 074 736	56
57	1653	63 365	2 732 409	125 678 141	6 021 318 633	296 720 109 245	14 926 096 567 929	57
58	1711	66 729	2 927 521	136 994 637	6 677 675 401	334 788 801 789	17 134 080 735 481	58
59	1770	70 210	3 132 900	149 111 998	7 392 599 700	376 969 335 430	19 622 732 220 300	59
60	1830	73 810	3 348 900	162 071 998	8 170 199 700	423 625 335 430	22 422 092 220 300	60
61	1891	77 531	3 575 881	175 917 839	9 014 796 001	475 145 709 791	25 564 835 056 321	61
62	1953	81 375	3 814 209	190 694 175	9 930 928 833	531 945 945 375	29 086 449 662 529	62
63	2016	85 344	4 064 256	206 447 136	10 923 365 376	594 469 447 584	33 025 430 301 696	63
64	2080	89 440	4 326 400	223 224 352	11 997 107 200	663 188 924 320	37 423 476 812 800	64
65	2145	93 665	4 601 025	241 074 977	13 157 397 825	738 607 814 945	42 325 704 703 425	65
66	2211	98 021	4 888 521	260 049 713	14 409 730 401	821 261 764 961	47 780 865 404 481	66
67	2278	102 510	5 189 284	280 200 834	15 759 855 508	911 720 147 130	53 841 577 009 804	67
68	2346	107 134	5 503 716	301 582 210	17 213 789 076	1 010 587 629 754	60 564 565 828 236	68
69	2415	111 895	5 832 225	324 249 331	18 777 820 425	1 118 505 792 835	68 010 919 080 825	69
70	2485	116 795	6 175 225	348 259 331	20 458 520 425	1 236 154 792 835	76 246 349 080 825	70
71	2556	121 836	6 533 136	373 671 012	22 262 749 776	1 364 255 076 756	85 341 469 239 216	71
72	2628	127 020	6 906 384	400 544 868	24 197 667 408	1 503 569 146 260	95 372 082 243 504	72
73	2701	132 349	7 295 401	428 943 109	26 270 739 001	1 654 903 372 549	106 419 480 762 601	73
74	2775	137 825	7 700 625	458 929 685	28 489 745 625	1 819 109 862 725	118 570 761 035 625	74
75	2850	143 450	8 122 500	490 570 310	30 862 792 500	1 997 088 378 350	131 919 149 707 500	75
76	2926	149 226	8 561 476	523 932 486	33 398 317 876	2 189 788 306 926	146 564 344 279 276	76
77	3003	155 155	9 018 009	559 085 527	36 105 102 033	2 398 210 687 015	162 612 867 546 129	77
78	3081	161 239	9 492 561	596 100 583	38 992 276 401	2 623 410 237 719	180 178 436 401 041	78
79	3160	167 480	9 985 600	635 050 664	42 069 332 800	2 866 497 743 240	199 382 345 387 200	79
80	3240	173 880	10 497 600	676 010 664	45 346 132 800	3 128 641 743 240	220 353 865 387 200	80
81	3321	180 441	11 029 041	719 057 385	48 832 917 201	3 411 071 279 721	243 230 657 842 161	81
82	3403	187 165	11 580 409	764 269 561	52 540 315 633	3 715 077 951 145	268 159 204 898 929	82
83	3486	194 054	12 152 196	811 727 882	56 479 356 276	4 042 018 324 514	295 295 255 883 556	83
84	3570	201 110	12 744 900	861 515 018	60 661 475 700	4 393 316 356 130	324 804 290 544 300	84
85	3655	208 335	13 359 025	913 715 643	65 098 528 825	4 770 465 871 755	356 861 999 372 425	85
86	3741	215 731	13 995 081	968 416 459	69 802 799 001	5 175 033 106 891	391 654 781 594 121	86
87	3828	223 300	14 653 584	1 025 706 220	74 787 008 208	5 608 659 307 900	429 380 261 081 904	87
88	3916	231 044	15 335 056	1 085 675 756	80 064 327 376	6 073 063 394 684	470 247 820 718 896	88
89	4005	238 965	16 040 025	1 148 417 997	85 648 386 825	6 570 044 685 645	514 479 155 614 425	89
90	4095	247 065	16 769 025	1 214 027 997	91 553 286 825	7 101 485 685 645	562 308 845 614 425	90
91	4186	255 346	17 522 596	1 282 602 958	97 793 608 276	7 669 354 937 686	613 984 947 550 156	91
92	4278	263 810	18 301 284	1 354 242 254	104 384 423 508	8 275 709 939 030	669 769 607 673 804	92
93	4371	272 459	19 105 641	1 429 047 455	111 341 307 201	8 922 700 122 479	729 939 694 734 561	93
94	4465	281 295	19 936 225	1 507 122 351	118 680 347 425	9 612 569 903 535	794 787 454 153 825	94
95	4560	290 320	20 793 600	1 588 572 976	126 418 156 800	10 347 661 794 160	864 621 183 763 200	95
96	4656	299 536	21 678 336	1 673 507 632	134 571 883 776	11 130 419 583 856	939 765 931 574 016	96
97	4753	308 945	22 591 009	1 762 036 913	143 159 224 033	11 963 391 588 785	1 020 564 216 052 129	97
98	4851	318 549	23 532 201	1 854 273 729	152 198 432 001	12 849 233 969 649	1 107 376 769 376 801	98
99	4950	328 350	24 502 500	1 950 333 330	161 708 332 500	13 790 714 119 050	1 200 583 304 167 500	99
100	5050	338 350	25 502 500	2 050 333 330	171 708 332 500	14 790 714 119 050	1 300 583 304 167 500	100

TABLE V. Sums of Logarithms of Integers

x	Σ lg x	Σ (x lg x)	Σ (lg x)³	x	Σ lg x	Σ (x lg x)	Σ (lg x)³
1	0,0000000	0,0000000	0,0000000	51	66,1906450	1982,0883971	93,3838763
2	0,3010300	0,6020600	0,0906191	52	67,9066484	2071,3205710	96,3285438
3	0,7781513	2,0334238	0,3182638	53	69,6309243	2162,7071920	99,3016711
4	1,3802112	4,4416637	0,6807400	54	71,3633180	2256,2564551	102,3028592
5	2,0791812	7,9365137	1,1692991	55	73,1036807	2351,9764030	105,3317215
6	2,8573325	12,6054212	1,7748184	56	74,8518687	2449,8749325	108,3878829
7	3,7024305	18,5211075	2,4890091	57	76,6077436	2549,9597993	111,4709794
8	4,6055205	25,7458274	3,3045806	58	78,3711716	2652,2386229	114,5806577
9	5,5597630	34,3340100	4,2151594	59	80,1420236	2756,7188916	117,7165745
10	6,5597630	44,3340100	5,2151594	60	81,9201748	2863,4079666	120,8783964
11	7,6011557	55,7893295	6,2996581	61	83,7055047	2972,3130866	124,0657990
12	8,6803370	68,7395045	7,4642903	62	85,4978964	3083,4413713	127,2784670
13	9,7942803	83,2207681	8,7051601	63	87,2972369	3196,7998259	130,5160934
14	10,9404084	99,2665606	10,0187696	64	89,1034169	3312,3953443	133,7783793
15	12,1164996	116,9079295	11,4019602	65	90,9163303	3430,2347124	137,0650341
16	13,3206196	136,1738492	12,8518651	66	92,7358742	3550,3246122	140,3757742
17	14,5510685	157,0914808	14,3658697	67	94,5619490	3672,6716240	143,7103234
18	15,8063410	179,6863859	15,9415788	68	96,3944579	3797,2322300	147,0684123
19	17,0850946	203,9827044	17,5767895	69	98,2333070	3924,1628173	150,4497786
20	18,3861246	230,0033043	19,2694686	70	100,0784050	4053,3196801	153,8541654
21	19,7083439	257,7699095	21,0177324	71	101,9296634	4184,7590228	157,2813229
22	21,0507666	287,3032084	22,8198311	72	103,7869959	4318,4869626	160,7310069
23	22,4124944	318,6229487	24,6741338	73	105,6503187	4454,5095314	164,2029790
24	23,7927057	351,7480185	26,5791169	74	107,5195505	4592,8326786	167,6970062
25	25,1906457	386,6965187	28,5333531	75	109,3946117	4733,4622734	171,2128609
26	26,6056190	423,4858257	30,5355027	76	111,2754253	4876,4041064	174,7503207
27	28,0369828	462,1326474	32,5843049	77	113,1619160	5021,6638922	178,3091670
28	29,4841408	502,6530722	34,6785713	78	115,0540106	5169,2472713	181,8891890
29	30,9465388	545,0626142	36,8171792	79	116,9516377	5319,1598115	185,4901776
30	32,4236601	589,3762518	38,9990664	80	118,8547277	5471,4070104	189,1119291
31	33,9150218	635,6081643	41,2232261	81	120,7632127	5625,9942970	192,7542442
32	35,4201717	683,7732636	43,4887026	82	122,6770266	5782,9270329	196,4169276
33	36,9386857	733,8842237	45,7945871	83	124,5961047	5942,2105145	200,0997884
34	38,4701646	785,9545068	48,1400148	84	126,5203840	6103,8499746	203,8026391
35	40,0142326	839,9968884	50,5241609	85	128,4498029	6267,8505832	207,5252965
36	41,5705351	896,0237784	52,9462384	86	130,3843013	6434,2174510	211,2675808
37	43,1387369	954,0472422	55,4054951	87	132,3238206	6602,9556260	215,0293157
38	44,7185205	1014,0790189	57,9012113	88	134,2683033	6771,0701012	218,8103286
39	46,3095851	1076,1305385	60,4326979	89	136,2176933	6947,5658118	222,6104500
40	47,9116451	1140,2129382	62,9992941	90	138,1719358	7123,4476376	226,4295137
41	49,5244289	1206,3370763	65,6003659	91	140,1309772	7301,7204043	230,2673568
42	51,1476782	1274,5135465	68,2353041	92	142,0947650	7482,3888844	234,1238194
43	52,7811467	1344,7526901	70,9035233	93	144,0632480	7665,4577986	237,9987445
44	54,4245993	1417,0646079	73,6044600	94	146,0363758	7850,9318169	241,8919781
45	56,0778119	1491,4591710	76,3375716	95	148,0140994	8038,8155594	245,8033687
46	57,7405697	1567,9460313	79,1023352	96	149,9963707	8229,1135977	249,7327590
47	59,4126676	1646,5346306	81,8982465	97	151,9831424	8421,8304560	253,6800209
48	61,0939088	1727,2342100	84,7248186	98	153,9743685	8616,9706114	257,6450022
49	62,7841049	1810,0533179	87,5815814	99	155,9700037	8814,5384957	261,6275620
50	64,4830749	1895,0023131	90,4680804	100	157,9700037	9014,5384957	265,6275620

LITERATURE CITED

C. A. Bula,"Theory and evaluation of central moments in two dimensions. Sheppard's corrections. The simpler method of Mitropol'skii," Revista Union Mat. Argentina 5:1-97 (1940). S. S. Wilks, Mathematical Reviews 2:231 (1941).

C. E. Dieulefait, Théorie de la corrélation, Rosario (1935).

A. K. Mitropol'skii, "On calculating the moments by method of sums," Izv. Leningr. Lesn. Inst. 36:207-250 (1928).

A. K. Mitropol'skii, "On establishing the correlation equations by Tchebycheff's method," Bulletin de l'Académie des Sciences de l'URSS. Série Mathématique 1:125-134 (1937).

A. K. Mitropol'skii, "On the multiple nonlinear correlation equations," Bulletin de l'Académie des Sciences de l'URSS. Série Mathématique 3:399-406 (1939). J. Neyman, Mathematical Reviews 1:345 (1940).

A. K. Mitropol'skii, "The ordinary correlation equations," Usp. Mat. Nauk, IY, 5(33):142-175 (1949).

A. K. Mitropol'skii, Technique of Statistical Computations, Gos. Izdat. Fiz.-Mat. Lit., Moscow (1961). R. G. Laha Mathematical Reviews 24(3A):324-325 (1962).